EUREKA!

EUREKA!

Scientific Breakthroughs that Changed the World

Leslie Alan Horvitz

John Wiley & Sons, Inc.

This book is printed on acid-free paper. ♾

Published by John Wiley & Sons, Inc., New York.
Published simultaneously in Canada

Library of Congress Cataloging-in-Publication Data
Horvitz, Leslie Alan.
Eureka! : scientific breakthroughs that changed the world / by Leslie Alan Horvitz.
p. cm.
Includes bibliographical references.
ISBN 0-471-40276-1 (cloth : alk. paper)
1. Discoveries in science. I. Title.

Q180.55.D57 H67 2001
509—dc21

2001046890

10 9 8 7 6 5 4 3 2 1

CONTENTS

EUREKA!

Introduction

A Sudden Flash of Light

Scientific progress comes in fits and starts—years of toil in a laboratory may prove fruitless but a contemplative walk in the countryside can yield an astonishing breakthrough. That's what the nineteenth-century French mathematician Henri Poincaré found out when previous efforts to solve a particularly thorny mathematical problem had come to naught. "One morning, walking on the bluff," he wrote, "the idea came to me, with . . . brevity, suddenness and immediate certainty. . . . Most striking at first is this appearance of sudden illumination, a manifest sign of long, unconscious prior work. The role of unconscious work in mathematical invention appears to me incontestable."

As mysterious and unpredictable as such bursts of creative insight are, their occurrence is frequent enough that the phenomenon has been well chronicled. Another French mathematician, Jacques Hadamard described the experience: "On being very abruptly awakened by an external noise, a solution long searched for appeared to me at once without the slightest instant of reflection on my part . . . and in a quite different direction from any of those which I had previously tried to follow." The German mathematician, Carl Friedrich Gauss, offers a similarly arresting account of how he solved a problem that had resisted him for four years: "As a sudden flash of light, the enigma was solved. . . . For my part I am unable to name the nature of the thread which connected what I previously knew with that which made my success possible."

Creativity, whether in mathematics, the sciences, or the arts, is a funny business. You can spend long years of struggle in fruitless effort, banging your head against a wall as if it were possible to force an idea to come, only to hit upon a solution all at once, out of the blue so to speak, in the course of an idle stroll. The twelve scientists who make their appearance in this book all experienced the kind of sudden illumination that Poincaré, Gauss, and Hadamard are referring to. Almost invariably their insights came about in a moment of distraction or else burst forth from their unconscious while they slept. Religious epiphanies, by most accounts, appear to represent a similar phenomenon, where inspiration seems to strike like a lightning bolt.

There seems no accounting for when lightning will strike (or whether it will at all). "I have had my solutions for a long time, but I do not yet know how I am to arrive at them," lamented Gauss. What he's saying is that he knows he knows the solution but it remains infuriatingly unreachable, hiding out in some dark corner of his brain. And even if lightning does strike, the revelation may produce a solution for which proof is impossible to come by. André-Marie Ampère, the early nineteenth-century French physicist, remarked:

> I gave a shout of joy. . . . It was seven years ago I proposed to myself a problem which I have not been able to solve directly, but for which I had found by chance a solution, and knew that it was correct, without being able to prove it. The matter often returned to my mind and I had sought

> twenty times unsuccessfully for this solution. For some days I had carried the idea about with me continually. At last, I do not know how, I found it, together with a large number of curious and new considerations concerning the theory of probability.

Sometimes, of course, genius outstrips technology, with the result that the idea may wither and die on the vine, derided and mocked—or worse, simply ignored—for years, only to be rediscovered when science catches up to it. When Alfred Wegener, the German astronomer and meteorologist, proposed the theory of continental drift in 1912—which states that hundreds of millions of years ago the continents had made up one great landmass that has since split apart—he provoked a storm of outrage from fellow geologists. That the theory was largely correct had to wait until the early 1960s—more than thirty years after Wegener's death—before the technology was available to substantiate it. But Wegener never lost faith in the theory. That's another thing the scientists in this book have in common: however intense their opposition, they remained convinced that they were correct. The proof of their theories, they assumed, would eventually be found. As one mathematician explained, "When you have satisfied yourself that the theorem is true, you start proving it." That could be called the aesthetic approach—something that looks good and feels right, strangely enough, has a good chance of being right, even if you can't immediately figure out how to convince anyone of its truth.

The development of technology also often prods the mind into new ways of thinking. Without the refinement of X-ray diffraction techniques at the end of World War II, biophyisicists James Watson and Francis Crick could have had the most brilliant theories in the world, but they still would never have gotten anywhere in their efforts to fathom the structure of the DNA molecule. But the catch phrase of the movie *Field of Dreams* is also apt in this context: "Build it and they will come." Invent an instrument that can do something earlier devices could not and some researcher somewhere will want to experiment with it. Take the invention in the eighteenth century of an enclosed chamber that allowed the researcher to heat up various substances in order to study the gases they emitted. Using this device, the eighteenth-century English chemist Joseph Priestley tested several different compounds, just to find out what would happen, with no idea as to what kind of gases he might produce. In the process he discovered oxygen.

There is, of course, no way to be sure that an epiphany will ever occur. Deserving scientists can struggle with a problem for a lifetime and never experience a flash of insight that will result in a solution. But it would be a mistake to think of the scientists discussed here as passive vessels, moved to action only after being seized by a burst of illumination. On the contrary, these scientists were preparing themselves for their eureka moment for years in advance, even if they didn't realize it at the time.

Fortune favors the prepared mind, said French chemist Louis Pasteur, and the French scientist Bernard Fontenelle observed: "These strokes of good fortune are only for those who play well!" But what kind of preparation exactly, you might ask?

For one, a prepared mind is a curious mind, open even to possibilities that experts consider outlandish, impractical, or simply foolish. Fortune seems to favor scientists who love to rummage around for ideas in places no one else would think to look—"the trash cans of science," as the French mathematician Benoit Mandelbrot (b. 1924) puts it. Almost without exception, these scientists were and are indifferent to boundaries and have little compunction about poaching on territory claimed by another discipline if it suits their purposes. These men are generalists, and as such they are impossible to pigeonhole. Wegener, for instance, was trained as an astronomer and meteorologist (he was also a champion balloonist), yet it was as a geological theorist that he made his mark. (Was it any wonder that geologists of the day resented him?) Friedrich Kekulé the German chemist, who unraveled the mysteries of how carbon molecules bond, studied to be an architect. Priestley considered the ministry his primary calling; science was something he did in his spare time. Watson, codiscoverer of the double-helix structure of DNA, had been an ornithologist before he turned to genetics. Mandelbrot, trained as a mathematician, tackled engineering and economic problems. These scientists all share a capacity to draw from many diverse—and seemingly unrelated—sources. (To come up with his theory of continental drift, Wegener amassed data from fields as varied as meteorology, seismology, paleontology, and zoology, as well as geology.) In other words, these men are all great synthesizers. "The sudden activation of an effective link between two concepts or percepts, at first unrelated, is a simple case of 'insight,' " noted the psychologist D. O. Hebb. "The insightful act is an excellent example of something that is not learned, but still depends on learning."

These scientists are also distinguished by their penchant for looking at something in a way that may seem counterintuitive. Even what may appear, on the surface, to be a mistake or a botched effort can have unexpected, and profitable, consequences, as illustrated by an invention of a scientist named Spencer Silver. Silver, who worked for the 3M Corporation, had developed a polymer adhesive that formed microscopic spheres instead of a uniform coating, with the result that it took years to set. Here, you'd assume, was a product that would seem to have no commercial viability. What use after all is an adhesive that takes forever to bind? Enter another 3M scientist, Arthur Fry. Fry was looking for a better bookmark for his church hymnal—a bookmark that wouldn't fall out, but that wouldn't damage its pages, either. That's when he thought of Silver's adhesive. Sure enough, it worked excellently as a bookmark: it stuck to the page when needed and was easily removed, leaving the paper unharmed. From this observation Fry came up with the idea for the Post-it. Conventional wisdom would say that all adhesives must be strong, but obviously that isn't true. It's just that it took a prepared mind to see how a weak adhesive could have an eminently practical use as well.

It is helpful to recall, too, that until the twentieth century, the boundaries between disciplines were far more blurred than they are today, and in some cases, disciplines that are now a standard part of any curriculum simply did not exist as such. That made it easier for scientists to roam far afield without worrying about whether they had the requisite credentials. Chemistry, for example, didn't come into its own until late in the nineteenth century. Moreover, some fields were so new—paleontology, for example—that practically anyone with enough curiosity and gumption could have the opportunity to make a significant contribution. No one, for instance, questioned Priestley when he decided to write a book about electricity, even though he wasn't an authority on the subject. That was because no one else was, either. In a sense, these scientists are explorers, venturing into little-known territory with unreliable maps, unafraid of risking humiliation or professional disgrace, and endowed with frightening self-confidence. They are also unafraid to ask questions—questions that sometimes seem so obvious that no one before thought them worth posing. For instance, just how long is the coast of Britain? And why is the length of the border between Spain and Portugal one length as measured by the Spanish and another as measured by the Portuguese, even though both nations use the same metric system? (See chapter 12.)

As explorers, most of the scientists profiled in this book are highly adaptable. If the path they had initially chosen turned out to be blocked, rather than give up, they would look for another. As a result, they might end up straying far from their intended destination. As so often happens, though, the destination that they had in mind to begin with was not the destination that they were meant to achieve. The English naturalist Charles Darwin did not set sail on the *Beagle* to find evidence that would account for the origin of species. He merely saw it as an opportunity for adventure; yet it was the observations that he made along the way that ultimately put him on the path that would lead him to propose the theory of evolution. The experience of American physicist Steven Weinberg, a Nobel Prize winner in physics, offers another telling example. He was trying to come up with particle descriptions that would explain the strong nuclear force (the force that binds atoms together), but was getting nowhere. "At some point in the fall of 1967, I think while driving to my office at MIT," he said, "it occurred to me that I had been applying the right ideas to the wrong problems." He realized suddenly that his particle descriptions were in fact correct, and that while they were irrelevant to the strong force, they could be applied perfectly to the weak nuclear force (the force that produces radioactivity) and electromagnetism. The prepared mind must be prepared to be surprised.

Serendipity as well plays a role in some of the discoveries described in the pages that follow. British bacteriologist Alexander Fleming goes on holiday one summer, neglecting to put away a bacterial culture he had been working on. In the lab below, a researcher, specializing in fungi, leaves the door open to the hallway because of the stifling heat. The spores from the fungi drift up into Fleming's lab and alight on the petri dish containing the culture. Fleming returns to find that something remarkable has happened to the contaminated culture that he could not have predicted. (See chapter 7). Or take another example: Joseph Priestley sits down one day to conduct an experiment on an unknown gas. A lit candle is nearby. To his surprise, when the gas is exposed to the flame, it burns more brightly and vigorously. (See chaper 1.) Yet if Fleming and Priestley hadn't been such acute observers, they might not have recognized the implications of these happy accidents, and the discoveries of penicillin and oxygen would have had to wait for another day. It does you no good if an opportunity is staring you in the face and you don't see it for what it is, or don't know how to take advantage of it. That is something only a prepared mind can do.

Finally, we come to what might be called the ripeness or zeitgeist theory. Or to put it another way: Why does a scientific breakthrough occur at one time and not at another? Clearly, a certain body of knowledge has to be acquired and disseminated beforehand: in other words, a foundation of data from observations and experiments has to be there to build upon. If Russian chemist Dmitry Mendeleyev had been doing his research in 1840, for example, rather than in the 1860s, he would not have been able to invent the periodic table for the simple reason that not enough elements were known, making it impossible to organize them according to their properties. Further, the problem has to be defined—not always as easy a task as it may seem. Why do a bowling ball and a feather fall at the same rate? What is the purpose of DNA? Why does the eastern coast of Africa seem to fit so well into the western coast of South America? Without a grasp on the problem, it's difficult to find a solution. (Though a solution sometimes does materialize and then the scientist has to go hunting for the problem to which it belongs.)

A scientific body of knowledge need not consist only of established facts. False starts, wrong turns, and discredited theories also have an instructive value. Once you know the mistakes your predecessors have made, you can avoid repeating them. But in many instances, the mistakes may suggest avenues that could have been profitably pursued but were not, for whatever reason. Many scientists have failed, not because they were moving in the wrong direction but rather because they didn't go far enough. Without an awareness of all the missteps and blind alleys engineers had taken in the past, for instance, American engineer Philo Farnsworth wouldn't have been able to develop a television system that worked where its precursors had failed.

That said, it does seem that some ideas manifest themselves at certain times and often to several scientists at once. It's almost as if there's something in the air—a germ of an idea—that you can reach up and grab if you are clever or intuitive enough. The notion that the time is "ripe" for an idea, be it the periodic table, continental drift, a theory of gravity, or fractal geometry, may not be possible to prove, but you could make a good case for it. Joseph Priestley wasn't the only scientist to "discover" oxygen. So did a Swedish chemist named Carl Wilhelm Scheele. Archibald Couper, a Scottish chemist, put forward a theory about the structure of carbon compounds that was uncannily like the one advanced by Friedrich Kekulé. Julius Lothar Meyer, a German chemist, came up with a periodic table practically identical to Mendeleyev's. And, in the best-known example of this kind of coincidence, the English botanist Alfred Russel Wallace proposed a theory of evolution that was little different from the one Darwin was working on. In each of these instances, the discoveries were made almost simultaneously.

Why, then, do we know so much about Priestley, Kekulé, Mendeleyev, and Darwin, and so little about Scheele, Couper, Meyer, and Wallace? The explanations vary. In some instances, it was a matter of luck; in others, an accident of geography. And it should go without saying that scientists who were self-promoters were more likely to gain a place in the history books than their more self-effacing rivals. Fortune, it seems, may well favor the person with a prepared mind and grant him or her the chance to make a great discovery, but fortune is not always so willing to award that person the credit he or she deserves for having done so.

CHAPTER 1

A Breath of Immoral Air

Joseph Priestley and the Discovery of Oxygen

Possibly no scientist has ever been more acutely aware of the role of chance in scientific discovery than Joseph Priestley, who stumbled on the existence of oxygen without having any real idea of what he was doing. The English chemist held that "more is owing to what we call *chance*, that is, philosophically speaking, to the observation of events arising from *unknown causes*, than to any proper design, or pre-conceived theory in this business." Even the experiment that earned him an exalted place in scientific annals was unplanned. He confessed:

> I know that I had no expectation of the real issue of it. For my own part, I will frankly acknowledge, that, at the commencement of the experiments . . . I was so far from having formed any hypothesis that led to the discoveries I made in pursuing them, that they would have appeared very improbable to me had I been told of them; and when the decisive facts did at length obtrude themselves upon my notice, it was very slowly, and with great hesitation, that I yielded to the evidence of my senses.

Even after making his historic discovery, however, he was unable to recognize its true significance. He still adhered to the received wisdom of his day that ordinary air—the air we breathe—became saturated with a substance called "phlogiston" when the air could no longer support combustion or life. Phlogiston was thought to be transferred during burning and respiration—a unifying idea in eighteenth-century chemistry. In an ironic twist in which the history of science abounds, Priestley had unwittingly paved the way for overthrowing the theory of phlogiston once and for all.

As a theoretical concept, phlogiston originated in the second half of the seventeenth century when an eighteenth-century German chemist named George Ernst Stahl theorized that when anything burned, its combustible part was given off to the air. This was the part that he called phlogiston, from the Greek word for "flammable." Plants, he maintained, absorbed phlogiston from the air. Because the rusting of metals appeared to be analogous to combustion, Stahl reasoned, this rusting process, too, must involve loss of phlogiston. By the same token, heating metallic oxides—called the calx—with charcoal restored phlogiston to them. So he concluded that the calx must be an element and the metal a compound. Although some chemists were seduced by the theory of phlogiston, it exerted only minor influence until the latter part of the eighteenth century, when it came under attack by the famous French chemist Antoine-Laurent Lavoisier. In theory, metals should lose phlogiston when heated, but in fact, according to advocates of the phlogiston theory, they gained weight. So what, Lavoisier wondered, had happened to the elusive phlogiston?

A more promising avenue of scientific research involving the chemical reaction of gases opened up in the early 1700s when the British physiologist Stephen Hales invented a pneumatic trough. For the first time chemists had a device to collect and measure the volume of gases released from various solids when they were heated. And since it was a closed system, the gases—which were called "airs"—remained uncontaminated by ordinary air. In 1756 the British chemist Joseph Black published his findings on the reactions of two salts—magnesium and calcium carbonates—when heated. What Black found was that they gave off a gas, leaving behind a residue of what he called "calcined magnesia," or lime (the oxides). This residue, when combined with sodium carbonate—an alkali—regenerated the original salts. The most significant finding to emerge from Black's work was that, contrary to what was previously believed, gases could in fact enter into a chemical reaction. Black's research set the stage for the identification of several gases as separate substances.

A decade later, another British scientist, Henry Cavendish, isolated hydrogen, which he called "flammable air." Hydrogen was only the third so-called air to be identified after ordinary air and carbon dioxide, which Black labeled "fixed air." It seemed that many airs were left to be discovered, a challenge that was eagerly taken up by Joseph Priestley, a minister and a church dissenter who incidentally also happened to be a brilliant chemist.

Joseph Priestley was born on March 13, 1733, in Yorkshire, England, the son of a wool merchant and his wife. Joseph was only six years old when his mother died, apparently worn out after giving birth to six children in as many years. Unable to care for his oldest child, his father decided to pack the boy off to live with his sister-in-law, Sarah Keighley, and her husband. The Keighleys were both Dissenters and as such were to have a lasting influence over Priestley's life. (Dissenters did not believe in the doctrines of the Anglican Church, the denomination to which most of England adhered. The Keighleys would be called Unitarians today.) There was no question that Priestley was a prodigy. By the age of 16 he had already mastered Greek, Latin, and Hebrew. Then he went on to teach himself French, Italian, and German. After his graduation from a college catering to Dissenters, he decided to try his hand at the ministry, but at the same time he was drawn to academia, taking up a position as a professor of languages at a school in Warrington. Until this point Priestley had displayed no interest in science.

What prompted Priestley to attend a series of lectures on practical chemistry given by a prominent surgeon in Warrington is uncertain. Clearly, though, they made an impression on him, sparking a fascination for all things scientific. There were only so many knowledgeable people to talk with about science in Warrington, and so he began to make annual pilgrimages to London. There he had the opportunity to meet some of the leading intellectual and scientific lights of his age, one of whom was Benjamin Franklin. He told Franklin that he wanted to write a book about electricity. That he knew little about the subject failed to daunt him. Evidently Franklin perceived in him a kindred spirit—someone whose audacity equaled his own—for he proceeded to encourage Priestley to pursue the project. Franklin's faith was amply rewarded. The result was *The History and Present State of Electricity*, published in 1767. It was breathtaking in scope, summing up just about everything that was known about electricity at the time. Priestley also offered original contributions, based on his own experiments, noting, for example, that when he electrified a hollow sphere there was no charge inside, anticipating the inverse square law of electrical attraction.

Yet it was his ministry, and not science, that Priestley considered his most important work. Invited to assume the ministry of Mill Hill Chapel in Leeds, he seized the opportunity eagerly, welcoming a chance to preside over a congregation he described as "liberal, friendly and harmonious." But his decision to relocate to Leeds, barely six miles from where he grew up, also turned out to be a felicitous choice for reasons that had nothing to do with religion.

Priestley found lodging in Leeds next door to a brewery. Others might have been repelled by the odor that hung over the place. Not Priestley. He became curious to learn more about the "air" that effervesced from vats of fermenting liquor. The air, he observed, was capable of extinguishing lighted chips of wood. Mixed with smoke, the air descended to the ground, an indication that it must be heavier than ordinary or common air. The air released by fermentation was actually carbon dioxide—what Black had dubbed "fixed air." While Priestley knew no more about airs than he had about electricity, he immediately embarked on learning about them. The only way to find out the properties of fixed air, he believed, was to experiment with it. By dissolving the air in water, he found, it made bubbles. The idea occurred to him that this fizzy water might be useful for making sparkling wines and that it might also have a therapeutic purpose as well, because, he believed mistakenly, it might prevent scurvy in sailors on long voyages. In the course of these experiments he succeeded in inventing soda water, making him, in effect, the father of the soft drink industry. His achievement was recognized by the Royal Society, which gave him a medal for it in 1773.

Encouraged by this initial success, he turned his attention to other "airs" given off when various substances were heated. As he had become more interested in chemistry, he had begun to reflect on the similarity between the processes of burning and respiration. Investigating the properties of airs offered him a chance to explore these processes in a laboratory setting. To carry out his experiments, he constructed a simple but ingenious apparatus, consisting of a trough full of mercury over which glass vessels could be inverted to collect the gas. (This device, the invention of which is credited by some historians to Henry Cavendish, had an advantage over the one invented by Stephen Hales, which relied on water. Priestley's trough allowed the experimenter to collect and measure water-soluble gases, which Hales's could not do.) The substance to be heated was placed in a second, smaller glass vessel within the larger one. Priestley would then proceed to heat the substance by focusing the sun's rays on it using a twelve-inch lens. Between 1767 and 1773 he discovered four new "airs," including nitric oxide (nitrous air), nitrogen dioxide (red nitrous vapor), nitrous oxide (diminished nitrous air, more commonly known to dental patients as "laughing gas"), and hydrogen chloride (marine acid air). His account of his gas experiments, "On Different Kinds of Air," published in his opus *Philosophical Transactions* in 1772, attracted the attention of another scientist across the English Channel who was to play a crucial role in Priestley's own scientific endeavors—Antoine-Laurent Lavoisier, the same French chemist who had made a stir in scientific circles by blasting the whole notion of phlogiston.

Over the next few years, Priestley discovered ammonia (alkaline air), sulfur dioxide (vitriolic acid air), silicon tetrafluoride (fluor acid air), nitrogen (also discovered by British chemist Daniel Rutherford in 1772), and a gas later identified as carbon monoxide. In these experiments Priestley observed that light was important for plant growth and that green plants gave off a substance he called "dephlogisticated air." His findings set the stage subsequently for the systematic work on photosynthesis begun in 1779 by the Dutch physician Jan Ingenhousz and by the Swiss cleric-naturalist Jean Senebier.

As if this were not enough, Priestley somehow found the time to make an original contribution to the understanding of optics with the publication in 1772 of *History and Present State of Dis-*

coveries Relating to Vision, Light, and Colours, a book that brought him an invitation to join Captain James Cook's second voyage of exploration (1773–1775) as an astronomer. He was forced to decline, however, because of conservative opposition to his Unitarian views.

In December 1772, Priestley found new employment at the country estate of William Petty, Earl of Shelburne, as his librarian, literary companion, and tutor to his two young sons. The terms were generous. Shelburne gave him the freedom to preach and write as he wished and to continue to investigate airs to his heart's content.

On the first day of August 1774, Priestley decided to see what would happen if he extracted air from mercurius calcinatus (red mercuric oxide). Why this substance in particular? Even the chemist himself was uncertain. In his account of the experiment, written several years later, he wrote: "I cannot, at this distance of time, recollect what it was that I had in view in making this experiment: but I know that I had no expectation of the real issue of it." The only reason he could think of was that, having done so many similar experiments, he was always prepared to do another. It was almost as if he had nothing better to do. But there was one unusual aspect to this particular experiment that he would only appreciate later. "If, however, I had not happened, for some other purpose, to have lighted a candle before me," he wrote, "I would probably never have made the trial; and the whole train of my future experiments relating to this kind of air might have been prevented." Here was where chance intruded itself into the equation.

He followed the same routine that he had established with his earlier experiments on airs, first showering the substance with sunlight, intensified by his lens, until it was sufficiently heated to give off a gas. He then added water to see if it would dissolve. It did not. "But what surprised me more than I can well express," he remarked, "was, that a candle burned in this air with a remarkably vigorous air."

Priestley had discovered an air—that much was obvious—but what kind of an air exactly? He had no idea. Eager to further investigate the mysterious gas, he chose to find out what would occur if he exposed it to a living creature. In March 1775 he introduced an adult mouse into a glass apparatus filled with air from mercurius calcinatus. He expected it to survive for no more than fifteen minutes, by which time the air would be depleted. By air he meant so-called common air, which both mice and men depend on to live. But in this air, he noted with astonishment, the mouse lived for half an hour. When Priestley removed the creature and held it to the fire, it at once revived, apparently none the worse for wear. "By this I was confirmed in my conclusion, that the air extracted from mercurius calcinatus . . . was, *at least, as good* as common air; but I did not conclude that it was any *better.*"

All the same, he concluded that "from the greater strength and vivacity of the flame of a candle," this pure air "might be peculiarly salutary in the lungs in certain medical cases." He offered a caveat: "But, perhaps, we may also infer from these experiments, that though pure dephlogisticated air might be very useful as medicine, it might not be proper for us in the usual healthy state of the body . . . so we might . . . *live out too fast,* and the animal powers be too soon exhausted in this kind of pure air." He even went so far as to wonder whether he had overstepped certain ethical limits by his discovery. "A moralist, at least, may say, that the air which nature has provided for us is as good as we deserve."

But moral considerations aside, Priestley knew that people

would be curious to know just how far he carried his experiment. "My reader will not wonder, that, after having ascertained the superior goodness of dephlogisticated air by mice living in it . . . I would have the curiosity to taste it myself." Priestly admitted that he had and then went on to describe what it felt like: "The feeling of it to my lungs was not sensibly different from that of common air; but I fancied that my breast felt peculiarly light and easy for sometime afterwards. Who can tell but that, in time, this pure air may become a fashionable article in luxury. Hitherto only two mice and myself have had the privilege of breathing it."

Priestley quickly realized that this gas was the component of common air that made combustion and animal respiration possible. However, he reasoned that combustible substances burned more energetically in this gas, and metals formed calxes (or oxides) more readily because all the phlogiston had been expelled. Common air, by contrast, was partially filled with phlogiston. Air in which things had burned became less able to support combustion because, it was thought, it was saturated with phlogiston. Accordingly Priestely called the gas in which a candle flame burned with such "splendour and heat" dephlogisticated air. In fact, what Priestley had discovered was oxygen.

Priestley wasn't alone in his discovery, though he did not know it at the time. In 1775, the same year that Priestley was carrying out his experiments with mice, Swedish pharmacist Carl Wilhelm Scheele (1742–1786) also isolated oxygen. Like Priestley, Scheele had no formal training in chemistry and studied the elements of science while apprenticed to an apothecary. An absence of a university education, however, was no obstacle to his discovering several compounds and elements, including chlorine and barium. And like Priestley, Scheele discovered oxygen by experimenting on various oxides. But occupying an unassuming position as the proprietor of a pharmacy in Köping, Sweden, he never achieved the recognition that history has accorded the British chemist.

In October 1774, Prietley accompanied Shelburne on a journey through Europe. When the two arrived in Paris, Priestley took the opportunity to meet Lavoisier to tell him about his new dephlogisticated air. The meeting was a watershed event for the future of chemistry. Lavoisier was quick to appreciate the significance of Priestley's experiments. In fact, he understood the theoretical implications of the discovery, which Priestley did not. Lavoisier was not a modest man: two years prior to Priestley's visit, in 1772, he declared that he was "destined to bring about a revolution in physics and chemistry." Indeed he was as good as his word. Unlike older scientists, he realized that common air, as Priestley called it, was not an "element," but a component of air for which he had been searching. Now Priestley had brought him the missing piece of the puzzle.

Lavoisier wasted no time in repeating Priestley's experiments, carrying out intensive investigations all through 1775–1780. The results convinced him that Priestley's air was the active "principle" of the atmosphere. In a series of brilliant experiments, Lavoisier showed that air contains 20 percent oxygen and that combustion is due to the combination of a combustible substance with oxygen. He additionally interpreted its role in respiration. It was Lavoisier who in 1789 gave the substance the name oxygen. He derived the name from the Greek for "acid former," in the belief that all acids contained it.

Where Lavoisier parted ways with Priestley was in his repudiation of the phlogiston theory. In place of the phlogiston theory, Lavoisier demonstrated that oxygen from the air combines with the components of the combustible substance to form oxides of the component elements. In his "Discours preliminaire," a preface to his major work, *Traité elémentaire de chimie* (Treatise of Chemical Elements, 1789), Lavoisier put forth his theory that substances that could not be decomposed by chemical means were *substances simples*—the basic elements out of which matter was composed. His conception of elements is little different from the way in which elements are regarded today. In the *Traite,* Lavoisier again anticipated modern science by declaring that matter cannot be either created or destroyed, but only altered and modified in chemical reactions. An equal quantity (no matter how greatly transformed) of the material remains before and after the operation. This theory—which would later find its echo in Einstein's theory about the equivalence of matter and energy—is known as the law of conservation of mass.

In 1783 Lavoisier published his complete overhaul of chemical theory (he even introduced a new nomenclature to replace one that still used alchemy as its basis). To drive home the point that phlogiston was dead and buried, his wife ceremoniously burned books propounding the discredited theory. Unfortunately, Lavoisier's luminous career in chemistry came to an abrupt and bloody end on the guillotine during the French Revolution: he was beheaded for having served the ousted monarchy as a tax collector.

The French Revolution was to make itself felt in Priestley's life as well. Even while he was exploring the properties of atmospheric gases, he continued to espouse his belief in human progress and perfectibility, gaining renown as a political theorist. His *Essay on the First Principles of Government, and on the Nature of Political, Civil, and Religious Liberty* (1769), emphasizing individual rights, inspired the English economist and utilitarian Jeremy Bentham to coin the famous phrase "the greatest happiness of the greatest number." In 1779, for reasons not entirely clear, Priestley left Shelburne's employ and settled in the industrial town of Birmingham as minister of the New Meeting congregation. While in Birmingham he continued to churn out a steady stream of books on religion and theology, sermons, tracts, and catechisms. Never one to shy away from controversy, he wrote a book that repudiated most of the fundamental doctrines of Christianity. It's possible that Priestley would have remained in Birmingham for the rest of his life, but in July 1791, on Bastille Day, a riot broke out—instigated by conservative elements opposed to freethinkers like Priestley—and in the ensuing rampage a mob destroyed Priestley's church, his home, and his laboratory. Priestley and his family barely survived a lynching, escaping the inflamed mob with only minutes to spare. While the incident is usually referred to as the Birmingham Riot, evidence suggests that the events were carefully orchestrated. All the victims were members of the dissenting religions and supporters of the French Revolution.

With the destruction of Priestley's library and the loss of his manuscripts, future biographers have been deprived of much of the material that could shed light on the chemist's life and work. A definitive biography of Priestley's early years exists (Robert Schofield's *The Enlightenment of Joseph Priestley: A Study of His Life and Work from 1733 to 1773,* published in 1998), but a comprehensive biography of his polymath in his later years is still to be written. Priestley may be simply too complex a figure to pin down. Here was a man, after all, who was equally at home in the laboratory, investigating electricity and discovering gases, as he was in the church, puzzling out the mysteries of the Book of Revelation.

Priestley's attempt to pick up the pieces and resume work in the aftermath of the riot proved fruitless. When efforts to rebuild his library and lab foundered, Priestley left England for good. In August 1793 the Priestleys set sail for America. A year later they decided to make their home on a farm in Northumberland, Pennsylvania. Although he seemed a bit lost without his laboratory and library, he declined the offer to take up a Unitarian ministry in Philadelphia, preferring to stay put with his family on his farm. But even though he chose a life of relative obscurity, his admirers didn't forget him. One of them was Benjamin Franklin, who wrote in a letter to one of the chemist's students, "Remember me affectionately to the honest heretic, Dr. Priestley." Priestley's change of circumstances did not, however, cause him to alter his routine (though he did give up wearing his wig). He continued to write, devoting most of his time to the completion of his theological works, particularly his *General History of the Christian Church*. And as always he continued to conduct scientific experiments. When, for example, Erasmus Darwin (grandfather of Charles) proposed a theory of spontaneous generation, which stated that organisms could arise from inorganic material like sand, Priestley began to experiment with algae in an effort to prove Darwin mistaken. Priestley also kept up with the work being done by a new generation of British chemists, like John Dalton (1766–1844), who developed the basis for modern atomic theory, and Humphry Davy (1778–1879), who, among other achievements, discovered that diamonds are made of carbon. Ironically, the last scientific paper Priestley was to publish—in June 1797—took the form of a pamphlet titled *The Doctrine of Phlogiston Established.* This was at a time when the entire scientific community had dismissed phlogiston as a myth. Even as Priestley lay dying seven years later, he still clung to his belief in the very theory that his discovery of oxygen had done so much to demolish.

CHAPTER 2

Epiphany at Clapham Road

Friedrich Kekulé and the Discovery of the Structure of Carbon Compounds

Of the elements that compelled the attention of early chemists, possibly the most significant was carbon. Carbon is one of the few elements known to humans since ancient times, discovered in the mists of prehistory by the first person who happened to dig a chunk of charcoal out of a smoldering fire.

Although carbon makes up only 0.032 percent of the Earth's crust, it is very widely distributed and forms a vast number of compounds—more compounds, in fact, than those of all the other elements combined. With oxygen and a metallic element, carbon forms many important carbonates, such as calcium carbonate (limestone) and sodium carbonate (soda). Certain active metals react with it to make industrially important carbides, such as silicon carbide (an abrasive known as carborundum), calcium carbide (used for producing acetylene gas), and tungsten carbide (an extremely hard substance used for rock drills and metalworking tools). Carbon is also crucial to life—all living organisms contain carbon; the human body is about 18 percent carbon by weight; in green plants carbon dioxide and water are combined through photosynthesis to form simple sugars (carbohydrates).

The study of carbon forms the basis of what we now know as organic chemistry. Yet how all these multiple compounds of carbon were formed and how they linked up with other elements and compounds remained mysteries to scientists of the mid–nineteenth century. Some chemists even went so far as to contend that the structure of molecules was essentially unknowable, since as soon as they experimented with them to find out their properties the resulting reactions would disturb the structure, so that it could never be determined.

Curiously, the scientist considered responsible for uncovering the mystery of carbon compounds was neither an especially proficient practical chemist nor a good teacher. Moreover, his most outstanding discoveries originated not from experiments he conducted in the laboratory but rather were inspired by a dream—two dreams to be precise. At least that's how the legend goes.

That certain substances can combine more easily with some chemicals than others, a concept taken for granted today, wasn't recognized to any great degree until the eighteenth century. Drawing on the knowledge that they gained from a series of experiments, chemists compiled elaborate tables to demonstrate the relative affinities when different chemicals were brought together. These tables gave them the ability to predict many chemical reactions before actually testing them in the laboratory. These advances, leading to the discovery of new metals and their compounds and reactions, resulted in the birth of what is known as the science of analytical chemistry. The leading light of this movement was the brilliant French chemist Antoine-Laurent Lavoisier (1743–1794). Lavoisier clarified the concept of an element as a simple substance that could not be broken down by chemical means, a theory he described in his 1789 *Traité elémentaire de chimie* (Treatise of Chemical Elements). In addition, he replaced the old system of chemical names (which was still based on alchemical usage) with the rational chemical nomenclature used today.

By the beginning of the nineteenth century the precision of analytical chemistry had improved to such an extent that chemists were able to show that simple compounds contained fixed and unvarying amounts of their constituent elements. In some cases, however, they found that more than one compound could be formed between the same elements. In 1858, in a classic example of serendipity at work, the German chemist Friedrich Wöhler mixed two salts—silver cyanate and ammonium chloride—sure that he was going to produce an inorganic compound called ammonium cyanate. That didn't happen at all. Instead he found he had an *organic* compound—urea. This startling discovery led him to conclude that atoms could be arranged into molecules in different ways. If they were arranged one way they would produce one type of compound, but an alteration of a few atoms could produce a compound with entirely different properties. Both ammonium cyanate and urea have the same type and number of atoms, but their structural arrangement has made one inorganic and the other organic. These observations set the stage for the chemical atomic theory of the English scientist John Dalton (1766–1844) in 1803.

When two elements combined, Dalton said, the resulting compound contained one atom of each element. This meant that water could be given a formula corresponding to HO—hydrogen and oxygen. Applying this principle to other compounds, he calculated the atomic weights of other elements and drew up a table of the relative atomic weights of all the then known elements. Although his theory was marred by many errors, the underlying principle—that a precise quantitative value could be assigned to the mass of each atom—was correct.

But there was a major flaw in Dalton's theory. He made no distinction between atoms and molecules. For one thing, there was no way to distinguish between the proposed formulas for water HO and H_2O_2—hydrogen peroxide, a colorless, syrupy liquid that is a strong oxidizing agent. And his theory was also incapable of explaining why the density of water vapor, with its assumed formula HO, was less than that of oxygen from which it was partly derived. The solution to these problems was found in 1811 by the Italian physicist Amedeo Avogadro (1776–1856). Avogadro suggested that it was necessary to make a distinction between molecules and atoms. When oxygen combines with hydrogen it does so as a pair of atoms. The two oxygen atoms are split, with each oxygen atom combining with two hydrogen atoms, resulting in two molecules of water, each with the molecular formula H_2O.

Unfortunately, Avogadro's ideas were overlooked for nearly fifty years, neglect that sowed considerable confusion among chemists trying to come up with accurate calculations in carrying out their experiments. Avogadro's hypotheses didn't come back into favor until 1860. How atoms are actually put together in molecules represents one of the most striking advances in modern chemistry, leading to the development of what is known as structural theory. The theory made possible the prediction and preparation of many new compounds, including important dyes, drugs, and explosives—in the process spurring the Industrial Revolution.

The scientist—and legendary dreamer—who would contribute so much to unraveling the surprising ways atoms bond together was born on September 7, 1829, in Darmstadt, Germany, to a family descended from a Czech line of a Bohemian aristocracy. Friedrich August Kekulé certainly didn't fit the image of a bookish scientific prodigy. By all accounts, Friedrich was an extroverted boy, with a quick wit, an affable manner, and a penchant for gymnastics. He even exhibited some talent for juggling and mimicry. His interest in botany and butterfly collecting was regarded by his parents as more of a boyish pastime than as harbingers of a future in science. His talent for sketching they took more seriously, though, which is why they felt that their son might one day design beautiful buildings. And perhaps he would have become an architect, had he not in 1847 attended a chemistry class at the University of Giessen taught by the great chemist Justus von Liebig (1803–1873). As a discipline, chemistry was not held in high regard at the time, and Kekulé's resolve to put aside architecture in favor of an obscure science was not welcomed by his family, at least initially. But Kekulé was persistent and eventually they came to accept his decision.

A career in chemistry in those years required a nomadic nature. In pursuit of an education (and the means with which to finance it), Kekulé first moved to Switzerland and then relocated to Paris before settling in London, where he found employment at St. Bartholomew's Hospital. During his English sojourn he befriended another former student of Liebig's named Alexander W. Williamson. At the time, Williamson was trying to classify organic compounds in terms of their atomic structure. It was a subject that intrigued Kekulé. The problem Williamson was contending with, though, was the same one that had bedeviled chemists for years: If, in chemical reactions all was flux, how could the fixed, or static, configurations of atoms with respect to one another ever be discovered? The way most chemists dealt with this vexing philosophical conundrum was simply to ignore it. But Kekulé could not leave it alone. The solution—at least the key to the solution—apparently occurred to him while on a London bus on his way home. He described his revelation, which, he said years later, occurred in the summer of 1854:

> One fine summer evening, I was returning by the last omnibus, "outside" as usual, through the deserted streets of the metropolis, which are at the other times full of life. I fell into a reverie and lo! The atoms were gamboling before my eyes. . . . I saw how, frequently, two smaller atoms united to form a pair, how a larger one embraced two smaller ones; how a still larger one kept hold of three or even four of the smaller; whilst the whole kept whirling in a giddy dance. I saw how the larger ones formed a chain, dragging the smaller ones after them. . . . The cry of the conductor "Clapham Road" awoke me from my dream, but I spent part of the night putting on paper at least sketches of these dream forms. Thus began the structure theory.

This reverie, Kekulé insisted, provided the inspiration for the idea of the carbon chain. Whether structure theory began at the Clapham Road stop, as Kekulé related it many years later, is open to question. According to William H. Brock, editor of *The Norton History of Chemistry* (1992), his claim, "while psychologically plausible, does not seem to fit well with his public pronouncements at the time or the deep conceptual transformation from taxonomy to constitution that it involved." The matter, he concludes, remains controversial. What is not in dispute is the significance of Kekulé's discovery, whatever its provenance. For one thing, it led to the theory that carbon atoms—twelve in all—are linked together to form long chains. This idea also opened the way to an understanding of aliphatic compounds, the large class of organic molecules consisting essentially of straight or branched chains of carbon atoms, including the alkanes, the alkenes, and the alkynes and their derivatives. (The other major class of organic molecules—the aromatic compounds, so called because they generally give off a pleasant aroma—have another structure entirely.)

Kekulé was the first chemist credited with the discovery of the existence of bonds between atoms. Bonds holding molecules together are formed by the sharing of one or more pairs of electrons by two atoms. The number of shared electron pairs indicates whether it will be a single or a double bond. In a single bond, only one pair of electrons is shared by the atoms, with one atom usually contributing one electron. These rules were understood for many molecules, but not for carbon. Kekulé postulated that carbon atoms were also able to form multiple bonds by sharing two or three pairs of electrons between two atoms, making for double and triple bonds as well as single ones. Kekulé also held that there were different types of bonds linking pairs of carbon atoms with the atoms of other elements, like hydrogen. Different types of bonds account for different types of compounds. Ethane, for example, is formed by a single bond between carbon and hydrogen, whereas ethylene requires a double bond between carbon and hydrogen.

In addition, Kekulé's theory also sought to explain that in substances containing carbon atoms, it was reasonable to assume that some of the bonds of each carbon atom are bonded to other atoms. Up until Kekulé's time it was thought that molecules had one central atom. Kekulé realized that a molecule of carbon was arranged in quite a different manner. With his fertile imagination, the chemist could visualize the bonds between the atoms, which he depicted as lumpy clouds, later dubbed "Kekulé's sausages." (These sausages confounded printers, who vastly preferred a linear layout, when setting them into type.) Carbon, Kekulé contended, was tetravalent. This means, first of all, that each carbon atom has four elements attached to it. A valence is a number representing the capacity of a single atom (or radical, which is an unpaired electron) to combine with other atoms (or radicals). The value is an expression of the number of electrons that an atom can give to or accept from another atom (or radical) during a chemical reaction. The theory of valences, one of the most important in chemistry, is the achievement of British chemist Edward Frankland (1825–1899), who determined that the atoms of one element can combine only with specific numbers of atoms of another element.

Kekulé's theory of the carbon structure introduced three basic ideas: that carbon atoms can combine with one another to form chains of any length and complexity; that the valency of carbon is always four (tetravalency); and that by studying the outcome of chemical reactions involving carbon it is possible to derive information about the element's structure.

What makes Kekulé's theory all the more remarkable is that he conceived it entirely in his mind without ever conducting any experiments in the lab. How he managed to perform these intellectual and imaginative leaps that swept aside centuries of conventional thinking about carbon is still something of a mystery. To some degree that's because Kekulé himself never publicly described how he came to his conclusions until he was already of advanced age. By then chemistry had taken enormous strides in understanding the structure of organic compounds so that it's possible Kekulé was retroactively presenting his earlier work as if he had been aware of experiments that were only carried out much later. He also seems to have discounted the part his youthful interests in botany, butterflies, and taxonomy played in his carbon theory. He didn't start out investigating carbon molecules in order to discover their structure, but rather to distinguish them in a way so that he could classify their various types. It wasn't until he had probed deeply enough into the mysteries of the carbon molecule that he came to understand the structural implications of his own work.

By 1859 Kekulé was back in Germany, teaching at the University of Heidelberg and elucidating his theory of carbon structure for publication. The result, *Lehrbuch der organischen Chemie,* is considered the first textbook on organic chemistry.

While influential in spreading his ideas, Kekulé never got around to actually completing the work. Bear in mind that Kekulé came up with his theory without the benefit of any analytic tools or experiments. His theory represents a spectacular triumph of intellect and imagination—and perhaps as well the inspiration of dreams. While it seems reasonable to believe that Kekulé's background in architecture influenced his later work in chemistry, some scientific historians point out that he was also deeply interested in botany and taxonomy, both of which emphasize classification of type based on an organism's structure and characteristics.

After Kekulé's breakthrough, progress in structural theory of chemical compounds began to accelerate. Ironically, chemistry as a field still didn't enjoy much respect. In Heidelberg, Kekulé was obliged to construct his own lecture room and laboratory, which he put together at home, using his own funds. Fortunately, when he married, he also gained a benefactor in the person of his brother-in-law.

As is often the case in science, discoveries sometimes come in pairs: both Charles Darwin (1809–1882) and Alfred Russel Wallace (1823–1913) derived the theory of evolution almost simultaneously, working independently. But usually only one person is given credit for the discovery while the other languishes in obscurity. This is exactly what happened with the discovery of tetravalency of carbon. It is highly possible that, while Kekulé arrived at it on his own, he might not have been the first to propose the mechanism underlying the bonding of carbon molecules. That honor may belong to a Scot by the name of Archibald Scott Couper (1831–1892). Like Kekulé, he didn't start out to become a chemist; it was only in his late twenties, while studying philosophy in Glasgow, that he became interested in the subject.

In Paris, Couper found work in the laboratory of Charles-Adolphe Wurtz, an eminent chemist interested in investigating the properties of ammonia. While at Wurtz's laboratory, Couper wrote a paper titled *On a New Chemical Theory,* which probably contained the first statement about tetravalence of carbon and its chain-forming ability. His training in philosophy might have played a part in his thinking, helping him to make the connection between the letters in words and the atoms in molecules. Unlike Kekulé, who had visualized bonds as resembling clouds or sausages, Couper conceived of them as straight dotted lines, a configuration that is still in use today. "Only one rational formula is possible for each compound," Couper wrote, "and when the general laws governing the dependence of chemical properties on chemical structure have been derived, this formula will express all of these properties."

Couper asked Wurtz to help him have his paper read at the French Academy of Sciences. The only way for this to happen was for the paper to be sponsored by a member of the Academy; since Wurtz wasn't a member he had to enlist someone who was. In May 1858, while Couper was waiting to learn the outcome of Wurtz's efforts, Kekulé's historic paper on tetravalency appeared in print, proposing the same idea. Whether justified or not, Couper was convinced that Wurtz had deliberately sabotaged him and stormed out of his laboratory, never to return. (Couper's paper was finally read to the Academy on June 14, 1858.) Couper subsequently suffered a nervous breakdown and was later hospitalized. The blow was so crushing that he never again was able to produce any scientific work. The importance of Couper's contribution wasn't recognized until the twentieth century, when the Scottish chemist was "rediscovered."*

The notions of tetravalency and different types of bonds quickly gained acceptance after Kekulé introduced his concept of the tetravalent carbon atom, but another carbon compound—benzene—remained as much a puzzle as ever. Credit for its discovery in illuminating gas made from whale oil in 1825 belongs to the English scientist Michael Faraday (1791–1867). It was given its name in 1845 by August Wilhelm von Hofmann, the German chemist who detected the substance in coal tar. Benzene is obtained chiefly from coke-oven gas, which yields various grades of benzol. For chemists, the structure of benzene had been of interest ever since Faraday had discovered it. While benzene was known to consist of six carbon atoms and six hydrogen atoms, chemists puzzled over how to arrange a chain of six carbon atoms with six hydrogen atoms while still preserving the tetravalence of carbon. Benzene, while clearly a carbon compound, did not fit the same profile that the aliphatic compounds did. Benzene represented a different kind of compound, which Kekulé labeled aromatic.

There were many difficulties in the investigation of benzene, not the least of which was that the presence of slight impurities

* Although it was Couper who had used the word "structure" in a linguistic analogy to refer to the order and arrangements of carbon molecules, it was a Russian chemist, Aleksandr Butlerov (1828–1886), who popularized the phrase "chemical structure" in referring to a particular arrangement of atoms within a molecule as a cause of its physical and chemical properties.

lowered melting points considerably, making the differentiation and identification of the melting points difficult. Although the exploitation of benzene in manufacturing lay in the future, the compound was already being used to some extent in the coal-tar and petroleum industries of the time. Kekulé and others were interested in applying the new structural theory to the compound in hope of finally puzzling out its structure. They weren't starting from scratch. Given that the empirical formula of benzene—C_6H_6—and the rules of carbon bonding were known, it is scarcely surprising that structural formulae were proposed before Kekulé's solution in 1865. Between 1858 and 1861, both Couper and another chemist, Josef Loschmidt, had done so but were unable to back their proposed structural models up with experimental evidence. Loschmidt used a circle to represent benzene, which was his way of indicating that the structure was mysterious and indeterminate. Kekulé dismissed his speculative structure as "*confusionsformeln,*" which needs no translation.

By this point Kekulé occupied the chair of chemistry at the University of Ghent in Belgium. In contrast to the frugal circumstances of the University of Heidelberg, Ghent was a scientific paradise. Not only was Kekulé's laboratory fully funded, but he was given the right to design it and have it equipped to his specifications. When he wasn't in his lab he was at work on his perennially unfinished chemistry textbook. He would sit up long after midnight preparing for his lectures. He seemed to have taken to heart the warning of his old teacher Liebig, who told him that one must ruin one's health to be successful at chemistry.

Then everything came to a sudden, catastrophic halt. In 1862, his beloved wife, Stephanie, died in childbirth. Kekulé fell into a depression and for the first time in his life found that he was incapable of doing any work. Two fallow years passed before he felt well enough to resume his research. And the first problem he turned his attention to was the riddle of the benzene structure.

In an experience that eerily recalled his vision of the carbon chain, Kekulé maintained that the solution came to him in a dream, which, in the words of the noted writer Arthur Koestler, "was possibly the most important dream in history since Joseph's seven fat and seven lean cows." Kekulé described the event:

> I was sitting writing on my textbook, but the work did not progress; my thoughts were elsewhere. I turned my chair to the fire and dozed. Again the atoms were gamboling before my eyes. This time the smaller groups kept modestly in the background. My mental eye, rendered more acute by the repeated visions of the kind, could now distinguish larger structures of manifold conformation; long rows sometimes more closely fitted together all twining and twisting in snake-like motion. But look! What was that? One of the snakes had seized hold of its own tail, and the form whirled mockingly before my eyes. As if by a flash of lightning I awoke; and this time also I spent the rest of the night in working out the consequences of the hypothesis.

The serpent biting its tail inspired Kekulé to come up with the vital clue he needed to determine the structure of benzene—a discovery that Koestler called "the most brilliant piece of prediction to be found in the whole range of organic chemistry." Put in a somewhat simplified manner, Kekulé's revolutionary proposal was that the molecules of certain important organic compounds are not open structures but closed chains or "rings" in the form of a hexagon with alternate single and double bonds—a form that resembles a snake swallowing its tail. One objection to

Kekulé's theory was that his hexagonal arrangement didn't display the properties normally associated with the double-bonded structure of benzene.

Critics also pointed out that his original conception failed to account for the arrangement of two isomers of benzene called dibromobenzene, one of which had a single and the other a double bond. (Isomers are molecular compounds that have the same formula but different structures.) But while there were two possible structures, only one form of dibromobenzene is actually produced. Undaunted by the objections, Kekulé postulated that there were indeed two chains—a double and a single—but that they were alternating in rapid oscillations within the molecule. In effect, while there were two structures, only one isomer, because of this cycling effect, was actually expressing itself in a "twining and twisting" snakelike motion that his dream had so vividly revealed to him. While still containing some flaws, subsequently addressed by scientific research, Kekulé's conception constituted the best representation of the benzene molecule in the nineteenth century and remains today the basis of our understanding of aromatic compounds.

Again there are some skeptics who doubt the veracity of Kekulé's account. While it is perfectly plausible that creative ideas can be inspired by dreams or visions, critics point out, there is no particular reason to necessarily put much stock in Kekulé's story. He spoke about his dream many years after the fact, stirring suspicion in some colleagues' minds that he wanted to ensure the acceptance of his theory over other claimants and felt that a good story would give him a decided advantage. Even Kekulé acknowledged that sooner or later the idea of a hexagonal structure would have occurred to some inventive scientist if he "tinkered" with the problem long enough. All the same, Kekulé's story stuck. It also contributed one of the most famous adages in scientific history. "Let us learn to dream, gentlemen," Kekulé declared to colleagues at a scientific gathering, making sure to add, "but let us also beware of publishing our dreams until they have been examined by the wakened mind."

It is clear that Kekulé couldn't possibly have proposed his hexagonal model of benzene without having already established the model of the carbon molecule, notwithstanding the fact that the two structures are distinctly different. All aromatics, Kekulé maintained, contain a six-carbon nucleus represented by a closed chain of alternating (or oscillating) single and double bonds.

Not every chemist greeted Kekulé's innovative structural theories with open arms. Some leading scientists of the day dismissed his ideas as nothing more than "a tissue of fantasies." One of Kekulé's principal detractors was the noted German chemist Hermann Kolbe. Ironically, Kolbe shared Kekulé's belief that the job of a chemist was to investigate the chemical composition of elements and compounds rather than to become bogged down in trying to classify them according to their properties. Yet he objected to Kekulé's conclusions all the same. By the late 1860s almost all chemists of any note had embraced structural theory. Only Kolbe held out. Atoms, he averred, could simply not be arranged "democratically"—that is to say, in a configuration where none would occupy a position of more importance than any other. No, Kolbe insisted, one atom, usually carbon or a hydrocarbon radical, had to be more central than its neighbors. Methyl, for instance, was a "commando"—the metaphor is Kolbe's—while carbon was the corporal in charge of three hydrogen "privates." He could rail all he wanted, but such a hierarchical (not to mention martial) scheme simply didn't apply to the aromatics. But Kolbe nurtured such an obsessive loathing for Kekulé that he refused to modify or abandon his discredited theory. He ended his days in a state of paranoia, convinced that he was a victim of a Prussian chemical and military conspiracy. With Kolbe's death in 1884, all criticism of Kekulé's theories came to an end.

Kekulé's contribution to chemistry transcended the identification of the structure of carbon and aromatic compounds. Chemical properties were now understood as arising from the internal structures of molecules, which could now be "seen" and "read" by chemists using the proper analytic and synthetic tools. The shape of a hexagon, it turned out, held the future of chemistry. It would, however, take another generation or two before chemists began to draw on the principles of physics to more fully grasp the combining properties of the substances they were studying.

For Kekulé the discovery of benzene's structure solidified his reputation, earning him renown as one of the world's most brilliant chemists. His findings were far from theoretical. The parent substance of a large class of chemical compounds, benzene turned out to have several important—and lucrative—industrial uses, including the manufacture of numerous plastics, dyes, detergents, and insecticides.

In 1867 Kekulé returned to Germany, initially to take up a position as chair of chemistry at the University of Bonn. In 1876 he married again—to his former housekeeper—and although they had three children, the match was an unhappy one. An attack of measles, also in 1876, impaired his health for the remainder of his life and seems to have adversely affected his scientific output as well. Nonetheless, his star remained undimmed. In 1890 Kekulé was honored by a celebration commemorating the twenty-fifth anniversary of the creation of benzene ring theory at the Deutsche Chemische Gesellschaft in Berlin.

In his Kekulé Memorial Lecture at the London Chemical Society in1898, a new generation of chemists paid him tribute. One of them, Francis Japp, called Kekulé's benzene theory the "most brilliant piece of scientific production to be found in the whole of organic chemistry." He went on to say that "three-fourths of modern organic chemistry is directly or indirectly the product of his theory."

Reflecting upon his life in a speech at Bonn in 1892, four years before his death, Kekulé attributed his success to both a preoccupation with architecture, which enabled him to think about the spatial relationships of groups of atoms, and to his extensive travels, which, he said, gave him the experience to sort out the good from the bad. An ability to dream seems to have helped, too.

CHAPTER 3

A Visionary from Siberia

Dmitry Mendeleyev and the Invention of the Periodic Table

The riddle had baffled thousands of scientists throughout the Western world for centuries: If species could be organized into categories based on their distinctive characteristics, as the eighteenth century taxonomist Carl Linnaeus had done, could a similar order be found for the chemical elements? And further, could this order be made to apply even to elements that had yet to be discovered?

Although elements such as gold, silver, tin, copper, lead, and mercury have been known since antiquity, the first *scientific* discovery of an element didn't occur until 1669, when German alchemist Hennig Brand (d.c. 1692) isolated phosphorus from a urine sample. The discoveries of arsenic and cobalt followed shortly. The tide of new discoveries continued unabated through the eighteenth and nineteenth centuries as scientists identified platinum, nickel, hydrogen, nitrogen, oxygen, chlorine, manganese, tungsten, chromium, molybdenum, and titanium. By the middle of the nineteenth century, chemists had amassed a vast body of knowledge about the properties of elements and their compounds. They had managed to identify sixty-three elements and predicted a few others that hadn't been isolated. They even knew their atomic weights—although in some instances they'd gotten the weights wrong. (Atomic weights are based on the number of protons and neutrons in the atom's nucleus; the atomic number is based only on the number of protons.) So now chemists were in a position that they wouldn't have been in several years earlier—1840, say—since they had enough elements in hand to attempt a rational arrangement of them. However, it was one thing to know that a rational organization of the elements was conceivable in principle, it was quite another to actually construct one.

The problem was that the elements had such decidedly different characteristics that it was hard to see whether they had very much in common at all. Some elements, like oxygen, hydrogen, chlorine, and nitrogen, were all gases; others, like mercury and bromine, were liquids under normal conditions; the rest were solids. There were some very hard metals, like platinum and iridium, and soft metals, like sodium and potassium. Lithium was a metal so light that it could float on water, whereas osmium was a metal twenty-two and a half times heavier than water. The metal mercury was not solid at all but a liquid. Gold, when exposed to air, never tarnished, but iron, on the other hand, rusted easily. Iodine simply sublimed and vanished into vapor. Some elements united with one atom of oxygen, others with two, three, or four atoms. A few, like potassium and fluorine, were too dangerous to handle without gloves. It was little wonder that scientists had come to grief trying to impose some kind of order. Clearly, only someone who was either very sure of himself or a fool would try to tackle a problem that had left any number of brilliant scientists flummoxed and embarrassed.

Attempts to devise a system of classification for the elements extend as far back as the early nineteenth century, when the English chemist and physicist John Dalton (1766–1844) advanced the theory that matter is composed of atoms of differing weights and that they also combine in simple ratios by weight. This theory, first proposed in 1803, is considered the foundation of modern physical science. Once chemists knew about atomic weights, they began to search for arithmetic connections between them with two goals in mind. For one thing, they were interested in finding out whether there was any likelihood that all elements were composed of a simple, common substance; for another, they wanted to see whether occasional similarities in their properties indicated similarities in structure. In 1817 the German chemist Johann Dobereiner (1780–1849) noticed that the atomic weight of strontium fell midway between the weights of calcium and barium, elements possessing similar chemical properties. In 1829, after discovering the halogen triad composed of chlorine, bromine, and iodine and the alkali metal triad composed of lithium, sodium, and potassium, he proposed that nature must be made up of triads of elements. In his theory, known as the law of triads, the middle element had properties that were an average of the other two members if their order was governed by their atomic weight. The significance of the law of triads escaped most chemists of the day partly because of the still-limited number of known elements and partly because of the chemists' inability to distinguish between atomic weights and molecular weights.

All the same, the new idea of triads became a popular area of study. Between 1829 and 1858 a number of scientists found that the types of chemical relationships Dobereiner had observed extended beyond the triad to larger groups. During this period fluorine was added to the halogen group. Oxygen, sulfur, selenium, and tellurium were grouped into another family, and nitrogen, phosphorus, arsenic, antimony, and bismuth were classified as yet another. Research was hampered, however, by the fact that accurate values of the elements were not always available.

If a periodic table is regarded as an ordering of the chemical elements demonstrating the similarity of chemical and physical properties, credit for the first periodic table probably should be given to a French geologist, A. E. Beguyer de Chancourtois. In 1862 de Chancourtois transcribed a list of the elements positioned on a cylinder in terms of increasing atomic weight. The cylinder was constructed so that closely related elements were lined up vertically. This led de Chancourtois to propose that "the properties of the elements are the properties of numbers." De Chancourtois was first to recognize that elemental properties recur with every seven elements—the recurrence is what is meant by periodicity. His chart had some major flaws, though, in that it included ions and compounds in addition to elements.

The next important development occurred in 1865, when an English chemist named John Newlands (1837–1898) made a valiant attempt to overcome the hurdles that earlier scientists had run into in establishing once and for all a relationship of the elements. In a paper titled the *Law of Octaves,* Newlands classified the fifty-six established elements into eleven groups based on similar physical properties, noting that many pairs of similar elements existed which differed by some multiple of eight in atomic weight. The law of octaves stated that every succeeding eighth element on the list displayed properties similar to the first. This observation prompted him to draw a comparison between the table of elements and the keyboard of a piano, whose eighty-eight keys are broken down into octaves, or periods, of eight. "The members of the same group of elements," he declared, "stand to each other in the same relation as the extremities of one or more octaves in music." His theory was greeted with derision. The whole idea of comparing chemical elements to a piano struck his colleagues as too preposterous to take seriously.

Newlands was onto something, though. Chemists in France,

Switzerland, and the United States had made similar observations in works published between 1860 and 1870. It was just that Newlands hadn't gone far enough. The task of corralling the elements into a coherent organization awaited someone with a fertile imagination and the courage to defy conventional wisdom. Thus was the stage set for a man who resembled an Old Testament prophet looking as though he had recently emerged from out of the wilderness to awaken the world from slumber. Only in this case the wilderness was Siberia and the prophet's name was Dmitry Mendeleyev.

No one who ever laid eyes on him could forget him; he was outsized in every way. Some said he bore an unsettling resemblance to Rasputin. As he grew older, personal appearance became less and less significant to him. "Every hair acted separate from the others," one startled observer wrote. He only cut his hair and trimmed his beard once a year. His slovenly appearance was misleading, though. He was a consummate and even obsessive scientist, a philosopher and a dreamer, but also a political agitator. While he believed it to be "the glory of God to conceal a thing"—the order of the elements among them—he was also convinced that it was "the honor of kings to search it out."

Dmitry Ivanovich Mendeleyev was born in Tobolsk, Siberia, on February 7, 1834, the youngest of fifteen children born to Maria Dmitrievna Komiliev and Ivan Pavlovich Mendeleyev. (The name Mendeleyev, alternatively spelled Mendeleev or Mendeléeff, roughly translates as "to make a deal.") When he was still young his father died of consumption, leaving his family without a kopeck. In spite of their sudden plunge into destitution, Dmitry's mother, a headstrong Tartar beauty, resolved to get him the best education possible in Russia. That meant the University of Moscow, not some ramshackle school in the provinces. After a journey of four thousand miles across the countryside, however, she and her son found themselves turned away by the university on the grounds that, as a Siberian, Dmitry was not eligible for admission. Undaunted, they proceeded to St. Petersburg, where Maria secured her favorite son a place at a gymnasium (secondary school). There he was expected to prepare to enter the prestigious University of St. Petersburg.

Even as an adolescent, Dmitry demonstrated the independent streak that would mark his scientific career. He wanted nothing to do with Latin or history—which were then thought necessary for a "classical education"—declaring them dead topics and a waste of his time. "We could live at the present day without a Plato," he would write many years later, "but a double number of Newtons is required to discover the secrets of nature, and to bring life into harmony with the laws of nature."

Her strength exhausted by her struggle to set Dmitry on the path she had planned for him, Maria died shortly after his acceptance at St. Petersburg University. "Refrain from illusions," she told Dmitry on her deathbed; "insist on work and not on words. Patiently search divine and scientific truth." He took her words to heart. During his third year at the university, Mendeleyev was stricken with an illness that was diagnosed as tuberculosis. He was given two years at most to live. But Mendeleyev had absolutely no intention of dying before achieving his life's ambitions. He was twenty-one years old, already driven by, as one writer put it, "the vision of the Russian people" whom he knew he could aid through science.

Defying his doctor's dire prognosis, Mendeleyev went on to make a full recovery and in 1856 was well enough to defend his master's thesis, *Research and Theories on Expansion of Substances Due to Heat.* He impressed his instructors so much that he was retained to lecture in inorganic chemistry. Since he could not find a textbook that suited his needs, he set about writing his own. The result was the classic *Principles of Chemistry.*

Mendeleyev's love affair with chemistry was by no means limited to theory. He was also a very practical man, interested in seeing science put to work in solving the problems of the world, whether it meant carrying out experiments to improve the yield and quality of crops, or coming up with ideas to modernize the Russian soda and petroleum industries. Never one to shy away from controversy, he castigated the U.S. petroleum industry while on an 1876 visit to the United States, contending that oil interests were more interested in expanding production than paying attention to efficiency or the quality of their products. At home he was equally critical of the way Russian oil resources were being exploited by foreign interests. Russia, he said, should develop its oil resources for its own profit.

Controversy also dogged his personal life. In 1882, he sought to extricate himself from a loveless marriage in order to marry his niece's best friend, Anna Ivanova Popova, who was many years his junior. Unfortunately, he hadn't reckoned on the opposition of the church. Under Russian Orthodox law, he was told, he would be unable to legally marry again for seven years. Never one to be deterred by the rules of any institution, he simply paid off a pliable Orthodox priest, who granted him a dispensation. (The priest was subsequently defrocked.) According to the Orthodox Church, though, Mendeleyev was still officially a bigamist. But by this time he was so famous that even the czar refused to be scandalized. "Mendeleyev has two wives. Yes, but I have only one Mendeleyev."

For more than thirteen years Mendeleyev had assiduously collected data on the sixty-three known elements from every conceivable source that he could find. Armed with this knowledge, he was determined to avoid the traps into which other scientists had fallen in their efforts to find a coherent pattern among the elements. He derided theories based on triads and hidden arithmetical series. In his view there had to be one factor that could account for both the similarities and the differences in the elements—including elements that remained to be discovered. Before he could start to assemble the pieces of the puzzle, though, he had to envision what the whole would look like when he was done. He had already laid the groundwork in his *Principles of Chemistry,* in which he organized his material around the families of the known elements that displayed similar properties. The first part of the text was devoted to the well-known chemistry of the halogens. Next, he covered the chemistry of the metallic elements in order of their combining power with oxygen—alkali metals first (combining power of one), alkaline earth metals second (power of two), and so on. However, he found it difficult to classify metals such as copper and mercury, which had multiple combining powers; sometimes the power was one and at other times it was two.

While trying to sort out this dilemma, Mendeleyev noticed patterns in the properties and atomic weights of halogens, alkali metals, and alkaline earth metals. Perhaps, he thought, it might be possible to extend this pattern to cover the other elements as well. He believed that even if he hadn't figured out the way all the elements were organized, at the very least he had found the common denominator that might make a solution possible someday. "There must be some bond or union between mass and the chemical elements," he wrote, "and as a mass of substance is ultimately expressed in the atom, a function dependence should exist and be discovered between the individual properties of the elements and their atomic weight. But nothing, from mushrooms to science can be discovered without looking and trying." Atomic weights, he was certain, could alone account for the similarities and differences in the elements.

The recognition that atomic weight was the one thing that all the elements had in common allowed Mendeleyev to proceed to the next step and begin to put the pieces of the puzzle together. But simply to arrange the elements according to their atomic weight, laying them out end to end, beginning with hydrogen (the lightest, with an atomic weight of 1) and finishing with uranium (then the heaviest, with a weight of 92), didn't satisfy him. It had been done before, and such an arrangement had failed to say anything useful about the elements' distinguishing characteristics. To make it easier to discern the pattern, he used sixty-three cards, one for each element. On each card he inscribed the element's symbol, its atomic weight, and its chemical and physical properties. Then he pinned the cards on the walls of his laboratory, arranging and rearranging them in hope of perceiving the organization of elements that God had concealed and that kings were meant to search out.

From the start, he knew that he was working at a disadvantage. No matter how much data he collected, he realized that there were still going to be huge gaps. But he was *never* going to have all the pieces for the simple reason that scientists hadn't found them all. In addition, he had to find a pattern that would hold not just for the sixty-three known elements but for countless others that were still to come. Better, he reasoned, to leave gaps for future scientists to complete than to corrupt the pattern by giving it a false sense of completeness.

Even when he traveled, Mendeleyev took his cards along. One can only imagine what his fellow train passengers must have thought as they watched this strange, unkempt man shuffling cards with obscure symbols and numbers on them. "The seed which ripens into vision may be a gift of the *gods,* but the labor of cultivating it so that it may bear nourishing fruit is the indispensable function of arduous scientific technique," wrote science writer Morris R. Cohen. For Mendeleyev, the long years of cultivation were about to pay off.

Like so many other scientists and inventors—including Thomas Edison and Friedrich Kekulé, who discovered in a dream the configuration of the benzene molecule—Mendeleyev had a habit of taking naps during the day. One afternoon in his office he suddenly woke from a dream, feeling strangely exhilarated. In a single stroke the dream had revealed to him practically the entire order of the elements. The date was February 17, 1668 (per the Julian calendar then in use in Russia) and Mendeleyev was 35 years old.

Mendeleyev hurriedly took down the cards from the walls and began to arrange them on a table, as if he were playing patience, his favorite game of solitaire. He arranged the elements into seven groups, starting with lithium (atomic weight 7), beryllium (9), boron (11), carbon (12), nitrogen (14), oxygen (16), and fluorine (18). The next element was sodium (21). Conveniently, sodium resembled lithium in terms of its physical and chemical properties. So it made sense to Mendeleyev to place sodium immediately below lithium in the table. Then he added five more elements until he came to chlorine, which had properties very similar to fluorine, under which it fell in turn. Everything dropping into place—the pieces made a perfect fit. In this manner he continued to arrange the rest of the elements. In a rectangular array, he was capturing similar elements with cousins laid out in columns, north to south. Working the other way, he was arranging elements that showed a gradual blend of properties from left to right. When he finished, he was startled to find that his arrangement had turned out to be so harmonious.

The very active metals—lithium, sodium, potassium, rubidium, and cesium—fell into one group (I). The extremely non-active metals—fluorine, chlorine, bromine, and iodine—all appeared in the seventh group. Since the experimentally determined atomic masses were not always accurate, Mendeleyev reordered the elements despite their accepted masses. For example, he changed the weight of beryllium from 14 to 9. This placed beryllium into group II above magnesium, whose properties it more closely resembled than nitrogen's, which it had been located above. In all, Mendeleyev found that seventeen elements had to be moved to new positions if he wanted their properties to correlate with those of other elements even though their accepted atomic weights indicated otherwise. These changes suggested to him that there were errors in the accepted atomic weights of some elements.

Mendeleyev had discovered that the properties of the elements were *periodic* functions of their atomic weights, as Newlands had hinted. To know the properties of one element in a group was to know the properties of all elements in the group. (Today the vertical columns are called groups and the horizontal rows are called periods.) What made Mendeleyev's predictions so astounding was that he had done no direct experiments to arrive at them. He had pulled his conclusions seemingly out of thin air.

But was Mendeleyev the true father of the periodic table? Or should he be obliged to share its paternity? In 1870 a German chemist named Julius Lothar Meyer (1830–1895) created a periodic table that was nearly identical to Mendeleyev's—another example of a major discovery made by two scientists working independently. It was almost as if the time was ripe for a periodic table. Meyer's 1864 textbook included a rather abbreviated version of a periodic table used to classify the elements. It consisted of about half of the known elements listed in order of their atomic weight and demonstrated periodic valence changes as a function of atomic weight. In 1868, Meyer constructed an extended table, which was the one similar to Mendeleyev's. But because Mendeleyev's table was published in 1869—a year before Meyer's—it was Mendeleyev who received the recognition for its creation. As often happens in such instances—a famous example is provided by the case of Darwin and Alfred Russel Wallace, who both arrived at the theory of evolution almost simultaneously—one scientist gets virtually all the credit and the other is remembered only by graduate students.

The advantage of Mendeleyev's table over previous attempts was that it exhibited similarities not only in small units such as the triads, but also in an entire network of vertical, horizontal, and diagonal relationships. What's more, all the elements in group I united with oxygen two atoms to one; all the atoms in the second group united with oxygen one atom to one, while those in the third joined oxygen two atoms to three. A similar pattern prevailed in the remaining groups as well. It was as if in solving one problem Mendeleyev had magically solved a second.

Astonishingly, the whole table worked so well that Mendeleyev couldn't help suspecting that it might be nothing but a coincidence. So he went back to check his calculations. And, in fact, it did seem as though he might have made an error: he had assigned platinum (with an atomic weight of 196.7) the position that gold (with an accepted atomic weight of 196.2) should have occupied. Naturally his critics were quick to jump on the discrepancy. But again Mendeleyev was proven right: when it was reevaluated, the atomic weight of gold turned out to be greater than that of platinum after all.

Other apparent flaws were pointed out: he had seemingly misplaced iodine, whose atomic weight was recorded as 127, and tellurium, 128. One of two scenarios was possible: either the new periodic table, or the atomic weight of tellurium, was wrong. With his characteristic self-confidence, Mendeleyev assured his detractors that the weight had to be off because his table was right. When chemists investigated the matter, they discovered that in fact tellurium belonged exactly where Mendeleyev had said it did; the recorded weight had been incorrect. (Mendeleyev did make some mistakes, proposing, for example, the existence of groups that were never found.)

If it worked as intended, the atomic weight system should be sufficiently elastic to predict new elements and even whole new groups, like the noble gases. On November 29, 1870, Mendeleyev decided to test the predictive power of the table by describing the properties of three undiscovered elements: eka-aluminum, eka-silicon, and eka-boron. He was audacious enough to specify their density, radius, and combining ratios with oxygen. Many of his fellow scientists were openly scornful. How could he be so certain that nature would oblige him?

It wasn't until November 1875 that Mendeleyev's predictions were first borne out. That was when a French chemist, François Lecoq de Boisbaudran (1838–1912), discovered one of the predicted elements (eka-aluminum, which was found in zinc ore mined in the Pyrenees). Just as Mendeleyev had forecast, the metal was easily fusible and could form alum. The element was named gallium. Then a German chemist, Clemens Winkler (1838–1904), set out to find the dirty-gray element with an atomic weight of about 72 and a density of 5.5 that Mendeleyev had also predicted. Winkler succeeded in his quest in 1886, turning up a grayish substance that had an atomic weight of 72.3 and a density of 5.5. In honor of his native land he named it germanium. The later discovery of eka-boron, the third predicted element, further confirmed Mendeleyev's honored place in the history of chemistry.

In spite of the fame that his scientific achievements brought him, Mendeleyev was still regarded with suspicion by the authorities for his progressive views and his advocacy of social reform. But he kept testing the limits—until he went too far. In 1890 Mendeleyev intervened on behalf of students at the university who were protesting unjust conditions. The minister of education forced his ouster from his post, asserting that he should have stuck to teaching and forgotten about politics. Police even went so far as to break up his final lecture out of fear that he might lead the students in an uprising. He quickly found another job, though, establishing a new system of import duties on heavy chemicals. Three years later he was appointed head of the Bureau of Weights and Measures, certainly an appropriate position for a man so obsessed with organizing the elements.

Neither domestic ferment nor political dissension could distract Mendeleyev from continuing his lifelong search for an underlying order of the elements. Curiously, he never intended to make a major contribution to chemistry; he was simply interested in resolving some of the chaos in the field for his students. After a while, though, the search took on a life of its own. As someone who was willing to make use of research that others had done in addition to his own work, Mendeleyev was one of the first modern scientists. Throughout his career he maintained an active correspondence with scientists around the world so that he could keep abreast of the latest developments in chemistry. Mendeleyev's achievement was possible only because enough was known about the elements to try to make some sense out of them.

A seeker to the end, Mendeleyev subsequently embarked on ever riskier scientific ventures that he might have been better off shunning. He spent years, for instance, trying to investigate the chemical composition of ether, which he believed was a material that belonged in his periodic table and which he was convinced consisted of particles a million times smaller than those of hydrogen. Here his vaunted powers of prognostication utterly failed him. The intangible something that he believed ether to be was in fact an intangible nothing.

However, his more quixotic pursuits never dimmed his fame nor did anything to tarnish his great achievement. By the time he died in 1907, Mendeleyev's table had expanded to include eighty-seven elements. At his funeral, students followed his bier holding aloft large reproductions of the periodic table. He was buried in St. Petersburg next to his beloved mother, to whom he owed so much.

There was still much to be learned about the legacy that Mendeleyev had left to the world. Not long before Mendeleyev's death, an English physicist named John William Strutt, Baron Rayleigh reported the discovery of a new gaseous element named argon which did not fit any of the known periodic groups. Odorless and invisible, and notoriously unsociable, in that it did not easily combine with any other element or compound, argon was one of a number of such inert gases. Actually argon had been discovered several years before—during a solar eclipse in 1868—but its significance wasn't appreciated for many years. The only evidence that there was something there at all was an orange-yellow line on a spectrograph. Mendeleyev had paid no attention to it and didn't bother to include it in his table. However, so commodious was the Russian's table that there was room for argon—and for the other so-called noble gases as well. In 1898, another British chemist, William Ramsay (1852–1916), suggested that argon did have a place in the periodic table between chlorine and potassium in a family with helium, despite the fact that argon's atomic weight was greater than that of potassium. This group was termed the "zero" group due to the zero valency of the elements, indicating that it had no combining power at all.

Although Mendeleyev's table demonstrated the periodic nature of the elements, it did not explain *why* the properties of elements recur periodically. For several years after the periodic table's publication, chemists struggled to explain the fact that chemical properties of the elements were a periodic function of their atomic weights. In 1913 the British physicist Henry Moseley (1887–1915) carried out experiments with X-ray diffraction to establish that the nuclear charge of an atom must indicate an element's position in the periodic table. With the discovery of isotopes of the elements, it became apparent that atomic *weight* was not the significant player in the periodic law as Mendeleyev and Meyer had proposed, bur rather that the properties of the elements varied periodically with atomic *number.**

The last major changes to the periodic table are largely due to the work of Glenn Seaborg, an American physicist who participated in the discovery of several new elements, beginning in 1940 with the discovery of plutonium, element 94. Elements continue to be found, with three new, though highly unstable, elements—116, 117, and 118—tentatively identified in 1999 alone. Scientists name new elements for countries (americium), for states (californium), and even for alma maters (berkelium). But mostly, they name the elements after illustrious scientists: einsteinium (Einstein), curium (Marie Curie), and fermium (Enrico Fermi). In1955 scientists identified element 101. In honor of the father of the periodic table, they called it mendelevium. The great Siberian visionary had finally earned a place at his own table.

---

* The *mass number* of an atom consists of the number of protons and neutrons in the nucleus. An *atomic number* is based only on the number of protons in the nucleus. *Isotopes* are different species of the same atom: they have the same atomic number but a different mass. An element's position in the periodic table is governed only by its atomic number, not its mass.

CHAPTER 4

The Birth of Amazing Discoveries

Isaac Newton and the Theory of Gravity

As a phenomenon, gravity—the attraction of one object for another—is the easiest fundamental force to observe. Gravity plays a crucial role in most processes on Earth. The ebb and flow of ocean tides, for instance, are caused by the exertion of the gravitational force on the Earth and the oceans by the Moon and the Sun. Gravity causes cold air to sink and displace less dense warm air, forcing the warm air to rise, so in that sense, gravity strongly influences the weather. Gravity also keeps the Earth intact by exerting what amounts to an inward pull that maintains the planet's integrity. That same inward pull is responsible for holding stars together. When a star's fuel runs down, the gravitational strength of the inward pull causes the star to collapse in on itself, a process which in some cases leads to the creation of a black hole. Here on Earth, gravity is what keeps every object—and all of us—from floating off into space.

Yet, while gravity may be an easy phenomenon to observe, it isn't very easy to explain. Some of the ablest minds in history were confounded when it came to accounting for why objects released from a certain height fall to the ground and stick rather than fly upward. In the fourth century B.C., the Greek philosopher Aristotle (384–322 B.C.) tackled the problem. He proposed that all things were made from some combination of the four elements, earth, air, fire, and water. If this were the case, he reasoned, then objects that were similar in nature would attract one another. So objects with more earth in them would be more attracted to other earth objects. Fire, on the other hand, would be repelled by earth, which explained why fire rose upward. In Aristotle's conception, the Earth was at the center of the universe, with the Moon, the Sun, planets, and stars revolving around it. He never attempted to draw any connection, however, between the force that kept bodies in motion in space and the force that caused objects to fall toward the Earth.

The challenge of trying to comprehend the explanation for gravity was next advanced by two extraordinary astronomers, the German Johannes Kepler (1571–1630) and the Italian Galileo Galilei (1564–1642). For both men, Aristotle's theory of gravity was unsatisfying. For one thing, it was based on a geocentric model—it put the Earth at the center of the universe—and for another, it failed to explain why objects remain on the Earth as it revolves on its axis, instead of flying off. And why, they wondered, do objects dropped from towers not fall to the west as the Earth is rotating to the east beneath them? And for that matter, how is it possible for the Earth, suspended in empty space, to rotate around the Sun without anything pushing it?

Galileo tried to bring logic to bear on the problem. Objects, he said, do not fly off the Earth because they are not really revolving rapidly. It doesn't matter how fast their apparent motion is, he argued, because compared to the Earth's revolutions, any object on the Earth is moving too slowly to gain the momentum to fly off. Galileo also had an answer as to why objects fall to the base of towers from which they are dropped, instead of being influenced by the direction of the planet's rotation. The falling object, he said, shared the same rotation of the Earth as the tower did. Bodies already in motion, he maintained, preserve that motion even when another motion is added to it. So, Galileo deduced, a ball dropped from the top of a mast of a moving ship would fall to the base of the mast. If the ball were allowed to move on a frictionless horizontal plane, on the other hand, it would continue to move unimpeded. This led Galileo to conclude that the planets, having been set in circular motion, continue in a circular orbit forever.

In making this assertion, Galileo studiously ignored the theory proposed by Kepler that the motion of bodies in space was not circular (as the sixteenth-century Polish astronomer Nicolaus Copernicus [1473–1543] had contended), but was, in fact, elliptical. To have agreed with Kepler, Galileo would have had to concede that his solution to the problem of gravity was mistaken.

According to Kepler, the planets didn't revolve about the Sun because they were set in perpetual motion, but were more likely kept in their orbits by some other force. But what was that force? Kepler thought that it must be the one force that appeared to be cosmic in nature, namely, magnetism. In 1600, William Gilbert (1836–1911), an English physicist, had carried out experiments that showed that the Earth functioned as a giant magnet. Given this fact, Kepler argued that magnetism must also emanate from the Sun, and that it was magnetism that propelled the planets around in their orbits. But once he'd gotten that far, Kepler was stumped. He couldn't resolve this vague idea into any coherent theory.

By the end of the first quarter of the seventeenth century, the

Aristotelian conception of gravity was all but dead. The trouble was that no one had come up with any satisfactory system to take its place. "The new philosophy calls all in doubt," complained one contemporary observer.

The next scientist to try his luck at the problem was the French philosopher and mathematician René Descartes (1596–1650), who never shied away from proposing theories meant to explain just about everything under the sun. All natural processes, said Descartes, could be explained by matter and motion. Although he offered a caveat that such mechanistic models were not the way nature probably worked, he contended that they had a certain utilitarian value as "likely stories," which he believed to be better than no explanation at all. With this model in mind, he contended, bodies once in motion tended to remain in motion in a straight line unless and until they were deflected by the impact of another body. Any changes of motion were always the result of such impacts. In Descartes's theory, a ball dropped from the top of a mast invariably dropped down at its base, moving in sync with the motion of the ship—unless another body hit it during its descent. He quickly ran into difficulty when he attempted to use this theory to explain why planets maintained an orbit around the Sun, forcing him to resort to a conception of the action of a "subtle matter" filling all space, which swept the planets up into a kind of whirlpool that kept them in motion. Aside from failing to determine the nature of this subtle matter, Descartes's model of the universe required the presence of a deity to keep the whole thing going. Voltaire observed that the Cartesian cosmos was like a watch that had been wound up at the creation and continued to tick away through all eternity.

The solution to the frustrating problem that had resisted the best efforts of Aristotle, Galileo, Kepler, and Descartes would have to wait for several years until a genius arrived who could figure it out for them. His name was Isaac Newton.

The man regarded as possibly the greatest scientific genius of all time, Isaac Newton was born into a family of yeoman farmers in Lincolnshire, England, on December 25, 1642—coincidentally the year of Galileo's death. (Newton's birth date is calculated according to the Gregorian calendar; under the old Julian calendar still in force in England at his birth, his birthday fell less memorably on January 4, 1643.) His formative years were not easy ones. His father died a few months before he was born, and Isaac himself was so sickly during his first year that his mother was surprised that he managed to live to his first birthday. Before he turned two, his mother remarried and moved out of Woolsthorpe, the family's manor house, to live nearby with her new husband. Young Isaac was left in the care of his maternal grandmother. He was only reunited with his mother in 1856, at the age of fourteen, when his stepfather died. The estrangement from his mother appears to have had a pronounced impact on the boy, making him secretive, isolated, aggressive, and overly sensitive to criticism.

It was with the hope of training him to manage her now much-enlarged estate that his mother took Isaac out of grammar school. But she soon realized that her son had little interest in herding cattle and harvesting crops. The boy preferred to bury his head in a book about mathematics. So his mother consented to putting him back into school. With the help of an uncle, he entered Trinity College at Cambridge University in June 1661. When he went off to Cambridge, the servants at Woolsthorpe Manor "rejoiced at his departure, deciding that he was fit for nothing but the 'Versity.' "

Even at the 'versity, however, Newton felt lonely and dejected.

Just as he did in childhood, he took refuge in books. "Plato is my friend," he wrote in one of his notebooks; "Aristotle is my friend, but my greatest friend is truth." Even though instruction at Cambridge was still dominated by the philosophy of Aristotle, students enjoyed some freedom of study by their third year. The less restrictive atmosphere allowed Newton to immerse himself in the new algebra and mechanical philosophy of Descartes, and in the Copernican-based astronomy of Galileo. At this stage Newton showed no great talent, and when he graduated with a bachelor's degree at the age of 25, he was considered an average student. However, there was one aspect of his personality that was apparent early on, and that was his tenacity. If a subject interested him, he attacked it doggedly and obsessively. To learn about the anatomy of the eye, for instance, he experimented on himself, once sticking a bodkin "betwixt my eye and the bone as near to the backside of my eye as I could."

Newton's scientific genius seemed to emerge fully formed in the summer of 1665, when the plague struck, forcing the university to shut down and compelling Newton to return to Lincolnshire. There he had ample time to think. It seems, in fact, that he did nothing *but* think. Within an astonishing period of only eighteen months, he embarked on the work that would revolutionize the understanding of mathematics, optics, astronomy—and, most important, gravity.

To appreciate Newton's revolutionary contribution, it might be useful to describe just how gravity operates. If an object held near the surface of the Earth is released, it will fall and pick up speed during its descent. This acceleration is caused by gravity, the force of attraction between the object and the Earth. The force of gravity on an object is equal to the object's weight, which is distinguished from the object's mass, or the amount of matter in the object. The weight of an object is equal to the mass of the object multiplied by the acceleration due to gravity.

For instance, a bowling ball that weighs 16 pounds is actually being pulled toward the Earth with a force of 16 pounds. In honor of the inventor of the universal law of gravity, scientists say that the bowling ball is being pulled toward the Earth with a force of 71 newtons (a metric unit of force abbreviated *N*). As we noted before, every object has gravity. This means that the bowling ball also pulls on the Earth with a force of 16 pounds (71 N), but because Earth is so massive the effect of the ball on Earth is imperceptible. In order to hold the bowling ball up and keep it from falling, a person must exert an upward force of 16 pounds (71 N) on the ball. This upward force acts to counter the 16 pounds (71 N) of downward weight force, leaving a total force of zero. The total force on an object determines the object's acceleration. In this case, the object is stationary, showing no acceleration whatsoever.

If the pull of gravity is the only force acting on an object, then all objects, regardless of their weight, size, or shape, will accelerate in the same manner. At the same place on the Earth, the 16-pound (71 N) bowling ball and a 500-pound (2,200 N) boulder will fall with the same rate of acceleration. As every second passes, each object will increase its downward speed by about 32 feet per second, resulting in an acceleration of 32 feet per second per second. In principle, a rock and a feather both would fall with this acceleration if there were no other forces acting on the two objects. In practice, however, air friction exerts a greater upward force on the falling feather than on the rock and makes the feather fall more slowly.

The mass of an object does not change as it moves from place to place. However, the acceleration due to gravity—and therefore the object's weight—will change because the strength of Earth's gravitational pull is not the same everywhere. The Earth's pull and the acceleration due to gravity decrease as an object moves farther away from the center of the Earth. At an altitude of 4,000 miles above the Earth's surface, for instance, the bowling ball that weighed 16 pounds (71 N) at the surface would weigh only about 4 pounds.

With the weight force reduced, the rate of acceleration of the bowling ball at that altitude would be only one quarter of the acceleration rate at the surface of Earth. The pull of gravity on an object also changes slightly with latitude. Because the Earth is not perfectly spherical, but bulges at the equator, the pull of gravity is about 0.5 percent stronger at the Earth's poles than at the equator.

The story of how Newton came up with his theory of gravity in a burst of inspiration while sitting under an apple tree seems almost too good to be true. But as it turns out, the story is more true than not. According to most accounts, he hit upon the grand theory of gravity—a single, comprehensive explanation as to how the force of gravity dictates the motion of the Moon and the planets—all at once. And while his epiphany did not come about because he was hit over the head by a falling apple, an apple tree apparently did play a crucial role in guiding the direction of his thoughts.

Newton's desk in his old bedroom looked out on an apple orchard, and it's entirely possible that at some point in his studies Newton must have gazed out on the garden. Further substantiation comes from Newton's friend, Dr. William Stokeley. In an account of a dinner he had with Newton in April 1726—sixty-one years after the fact—Stokeley wrote:

> After dinner, the weather being warm, we went into the garden and drank tea, under the shade of some apple trees,

> only he (Newton) and myself. Amidst other discourse, he told me he was just in the same situation as occasioned by the fall of an apple, as he sat in a contemplative mood. Why should that apple always descend perpendicularly to the ground, thought he to himself? Why should it not go sideways or upwards, but constantly to the earth's center? Assuredly, the reason is, that the earth draws it. There must be a drawing power in matter: and the sum of the drawing power in the matter of the earth must be in the earth's center, not in any side of the earth.

The apple always fell perpendicularly—or toward the center. Newton realized that there must be some kind of measurable relationship between the apple and the Earth, in other words, that both objects—apple and Earth—exerted gravitational force. As he elaborated his theory, he expanded its application to the relationship between bodies throughout the entire universe. As always, he was guided in his thinking by the same method he brought to bear on every problem: "from the phenomena of motions to investigate the forces of nature, and then from these forces to demonstrate the other phenomena."

Based on that premise, Newton could now begin to apply the property of gravity, evident in the fall of an apple to the Earth, to the motion of the Earth and of the planets and stars. He now had a theory to account for what "kept the planets from falling upon one another, or dropping all together into one center," in Stokeley's words. "This was the birth of these amazing discoveries, whereby he built philosophy on a solid foundation, to the astonishment of all Europe."

Newton's law of gravity is set out in the *Philosophiae Naturalis Principia Mathematica* (Mathematical Principles of Natural Philosophy, usually called the *Principia*), which appeared in 1687, and is still considered one of the most magisterial scientific works of all time. The law of gravity states that every particle of matter in the universe attracts every other particle with a force that varies according to its mass and to the inverse square of the distance between them. "I deduced that the forces which keep the planets in their Orbs must (be) reciprocally as the squares of their distances from the centers about which they revolve," he wrote.

In order to determine the mathematical relationship between the planets, Newton also needed to think about centrifugal force of a body moving uniformly in a circular path. In effect, centrifugal force is the opposite of gravity, since it is the force that pushes away from the center. While Kepler had derived the empirical laws to describe planetary motion, he had made no attempt to understand or define the underlying physical processes governing their motion. This was where Newton's theory represented a significant development.

Newton's great insight of 1665 was to imagine that the Earth's gravity extended to the Moon, counterbalancing the Moon's centrifugal force as it orbited the Earth. From his law of centrifugal force, Newton deduced that the centrifugal force of the Moon or of any planet must decrease as the inverse square of its distance from the center of its motion. For example, if the distance is doubled, the force becomes one-fourth as much; if the distance is trebled, the force becomes one-ninth as much. At the same time, the motion of the planets around the Sun is due to the same inverse-square law of gravity, where the force between two masses is inversely proportional to the square of the distance between them and proportional to the product of the two masses. In other words, when a planet is closer to the Sun, it will feel a stronger force accelerating it and hence will travel faster. By the same token, the attraction of the planets by the Sun is the same as the gravitational force attracting objects to the Earth. Newton further concluded that the force of attraction was proportional to the product of their masses. Newton claimed that this inverse-square law (law of gravity) is universal, in that it applies to any mass of material in the universe, whether it is an apple falling to the ground or a planet orbiting the Sun.

Given the law of gravity and the laws of motion, Newton was now in a position to explain a wide range of phenomena, such as the eccentric orbits of comets, the causes of the tides and their major variations, the precession of the Earth's axis (its wobbly gyration as it rotates), and the perturbation of the motion of the Moon by the gravity of the Sun. Virtually in one fell swoop, Newton had managed to impose an order on most of the known problems of astronomy and terrestrial physics, uniting in one coherent scientific theory the works of Galileo, Copernicus, and Kepler. The new Copernican worldview finally had a firm basis in reality. (It should be noted that during this time Newton wasn't solely concerned with gravity but was also investigating the nature of light. Using a prism of glass, he split white light into a range of colors, which he named the spectrum.)

Newton's theory of gravity still had some limitations: for instance, he had yet to achieve a correct understanding of the mechanics of circular motion, which he considered the result of a balance between two forces—one centrifugal, the other centripetal (toward the center)—rather than as the result of one force, a centripetal force that constantly deflects the body away from its inertial path, which is a straight line. And, too, he still believed that the Earth's gravity and the motions of the planets might be caused by the action of whirlpools, or vortices, as Descartes had claimed.

Several years passed during which Newton turned to other pursuits. In the early 1670s he immersed himself in the investigation of light and optics; in 1672 he was elected to the Royal Society for inventing the reflector telescope, which uses a combination of mirrors rather than lenses to focus light. This type of telescope, which is still used today, is known as the Newtonian reflector. Then, stung by an attack against his theory of light by a philosopher named Robert Hooke, Newton retreated from public view, occupying himself with work on chemistry and alchemy. (Newton was still a man of his time and alchemy was considered a legitimate field of study.) It wasn't until more than a decade later, in August 1684, that his interest in gravity was revived by a visit from the prominent British astronomer Edmond Halley (1656–1742).

Halley was anxious to learn whether Newton had had any success in solving a problem that had long perplexed him as well as virtually every other astronomer and scientist in England. The problem was this: Kepler had proven that planets move in elliptical orbits around the Sun. How then could these orbits be the logical consequence of an inverse-square law of gravity? While several scientists had speculated that this was the case, no one could prove it. The challenge was made even more daunting because the mathematics of the day was not advanced enough to solve the problem. Halley wasn't really very hopeful that Newton would have figured out a solution, either.

To Halley's amazement, however, Newton said that he had proven that elliptical orbits were a direct consequence of the inverse-square law of gravity. Newton had succeeded where so many others had failed in solving the problem of the elliptical orbits because in 1665–1666 he had conceived of a new form of mathematics to do so. His new mathematics would later be called calculus. (Differential calculus was invented independently by the German mathematician Gottfried Leibniz in 1675.) The "method of fluxions," as Newton termed his new mathematical tool, was based on his crucial insight that the integration of a function (or finding the area under its curve) is the mirror image of differentiating it (or finding the slope of the curve at any point). By using differentiation, Newton derived simple analytical methods that unified several disparate mathematical techniques previously applied on a piecemeal basis to such problems as finding areas, the lengths of curves, and tangents. Even though Newton could not fully justify his methods—rigorous logical foundations for calculus weren't developed until the nineteenth century—he is still credited with the invention of calculus.

Intrigued by Newton's announcement, Halley asked to see the proof. Obligingly, Newton began to rummage around in his files but failed to find it. Somehow he had managed to misplace one of the most important discoveries in the history of science. Halley urged Newton to go back and repeat the work. With Halley's encouragement and frequent prodding, Newton proceeded to do so. Three months after his visit to Newton, the astronomer received a short tract titled *De Motu* (On Motion) that contained the germ of the *Principia*, which Newton was still working on. It wasn't that the tract was so significant in and of itself—it said nothing, for instance, about the law of gravity. Its importance lay in the inspiration it provided Newton when he set about to revise and expand upon the theories in the tract regarding planetary dynamics. In the course of revising *De Motu*, Newton derived his famous three laws of motion:

I Every body preserves in its state of rest or uniform motion in a straight line except in so far as it is compelled to change that state by forces acted upon it.

II The rate of change of linear momentum is proportional

to the force applied, and takes place in the straight line in which that force acts.

III For every action there is an equally opposed reaction.

Using these three laws together with Kepler's laws of planetary motion, Newton explained the motion of the planets around the Sun as being due to the inverse-square law of gravity, where the force between two masses is inversely proportional to the square of the distance between them and to the product of the two masses. Newton found that the centripetal force holding the planets in their given orbits about the Sun must decrease with the square of the planets' distances from the Sun. So, when a planet is closer to the Sun, it will feel a stronger force accelerating it and consequently will travel faster; when the planet is farther from the Sun, it will feel less of a pull and travel slower. To describe the phenomenon Newton used the ancient Latin word *gravitas* (literally "heaviness" or "weight").

Even after completing the *Principia*, Newton was reluctant to have it published. He didn't feel any need to reach out to either the scientific community or the public. "I see not what there is desirable in public esteem were I able to acquire and maintain it," he wrote. But Halley continued to press him until Newton consented to allow the astronomer to present the work to the Royal Society, the ultimate arbiter of scientific achievement. As it happened, though, the Society had just published a book about fish that had bankrupted its treasury. The Society was so impoverished in fact that it paid its officers' salaries in copies of the fish book. Rather than let the *Principia* languish in obscurity, Halley scraped together the necessary funds out of his own pocket and paid for its publication. The scientific classic finally appeared in 1687. Within the framework of an infinite, homogeneous, three-dimensional, empty space and a uniformly and eternally flowing "absolute" time, Newton had in the *Principia* fully analyzed the motion of bodies acting under the influence of centripetal forces.

The publication of the *Principia* was not universally acclaimed. When the Royal Society received the completed manuscript of Book I in 1686, Newton's most aggressive detractor, the philosopher Robert Hooke (who had disputed Newton's work on optics), claimed that Newton had plagiarized his ideas. Hooke's charge is not entirely without merit. Hooke is credited with suggesting to Newton that since the planets move in ellipses with the Sun at one focus (Kepler's first law), the centripetal force drawing them to the Sun should vary as the inverse square of their distances from it. However, Hooke could never prove this theory mathematically, although he boasted that he could. Most scientific historians do not give any credence to his allegation. However, it is very likely that Newton could have assuaged Hooke, who was near death at this point, by graciously acknowledging his contribution. Instead, Newton methodically went through his manuscript and eliminated nearly every reference to Hooke he could find.

Objections from other quarters were raised as well. Because Newton repeatedly used the term "attraction" to describe gravitational force in the *Principia,* mechanical philosophers attacked him for reintroducing into science the idea that somehow matter alone could act at a distance upon other matter. Newton tried to deflect this criticism by maintaining that he had only intended to establish the existence of gravitational attraction and to discover its mathematical law. It was not his purpose to inquire into its cause. He was no more ready than his critics to believe that matter could act at a distance. But the more he thought about it, the more he decided that the notion of Cartesian vortices was not going to offer a satisfactory explanation for how gravity could act over large distances, either. However, that still left him with the problem of coming up with an alternative. In the early 1700s he reverted to the idea that some sort of material medium, or ether, must account for the existence of gravity. This ether, Newton surmised, had to be extremely rare so that it would not obstruct the motions of the planets, and yet it would also have to be very elastic or springy so that it could push large masses toward one another. To meet these criteria the ether would consist of particles endowed with very powerful short-range repulsive forces.

With the publication of the *Principia,* Newton was recognized as the leading natural philosopher of the age, but his creative career was effectively over. After suffering a nervous breakdown in 1693, he retired from research to seek a government position in London. In 1695 he became warden of the Royal Mint and in 1699 its master, an extremely lucrative position. Apparently he was as dogged in his pursuit of counterfeiters as he had been in ferreting out the laws of physics. In 1703 he was elected president of the Royal Society and was reelected each year until his death in 1727.

Intriguingly, for all his scientific genius, Newton was not fully convinced that his three laws of motion and his principle of universal gravity were sufficient to explain the universe if God weren't taken into account. Gravity, Newton more than once hinted, was a consequence of direct divine action, as were all forces for order and vitality. Absolute space, for Newton, was essential, because he regarded space as the "sensorium of God," and the divine abode must necessarily be the ultimate coordinate system. From his analysis he concluded that the mutual perturbations of the planets caused by their individual gravitational fields predicted the natural collapse of the solar system—unless God acted to maintain the cosmic order. "Absolute Space, in its own nature, without regard to anything external, remains always similar and immovable," he wrote in the *Principia*. "Absolute, True and Mathematical Time, of itself, and from its own nature, flows equably without regard to anything external."

Newton's law of gravity has held up remarkably well. However, as we now know, the universe is a much more complicated place than Newton could possibly have imagined. There is no uniformly and eternally flowing "absolute" time, as he maintained, no more than there is any such thing as "absolute" space, as we'll see in the next chapter. So, too, his theory of gravity, while providing a useful explanation for why apples fall to the ground, planets remain in their orbits, and humans don't fly off the surface of the Earth, runs into serious problems when it is applied to such phenomena as quasars and black holes. However, more than two centuries would have to pass before another scientist emerged who was equal to the task of unraveling the mysteries of gravity that Newton had failed to solve.

CHAPTER 5

The Happiest Thought

Albert Einstein and the Theory of Gravity

"Newton, forgive me," Albert Einstein (1879–1955) wrote in his *Autobiographical Notes*. "You found the only way which, in your age, was just about possible for a man of highest thought and creative power." And what Isaac Newton (1642–1727) found was indeed extraordinary. He single-handedly created modern physics, having developed a system that described the behavior of the entire cosmos. While others had done so before him, Newton was the first to advance theories that were based on mathematics and that made it possible for specific predictions to be confirmed by experiments in the real world. "There is only one universe to discover," noted a contemporary of the English polymath, adding, whether in admiration or disappointment, "and he discovered it."

The one drawback to this view, as Einstein recognized, was that Newton hadn't discovered *all* of it. Where Newton's theory of gravity is concerned, for instance, it remains one of the most significant scientific triumphs of all time. Its predictions were vindicated by observations with extraordinary accuracy. When astronomers in England and France used the motion of Uranus to predict the existence of Neptune, they relied on principles established by Newtonian gravity. In that respect, the theory served as a very good approximation to reality. However, as with any theory, there were special situations in which it did not apply. There were two cases where scientists who followed Newton found to their puzzlement and vexation that the theory seemed seriously flawed.

First, it gave the wrong prediction for a specfic part of Mercury's orbit, which turned out to be faster than Newton's theory would allow. Mercury's orbit is elliptical, as predicted by Newton's theory of gravity, but the ellipse doesn't stay in precisely the same place all the time. (It was Johannes Kepler [1571–1630] who had first postulated that planetary orbits were elliptical and not circular as previously thought). Most of this precession could be understood in the context of Newton's theory of gravity by taking into account perturbations of the orbit due to the gravitational pull of other planets. However, there still remained a discrepancy between the prediction and the observed value of about 40 seconds of arc per century. Why this discrepancy should exist was a complete mystery to scientists at the beginning of the twentieth century. They even went so far as to postulate the existence of an unseen planet, Vulcan, on the far side of the Sun in order to explain it.

Second, Newton's theory did not explain why the gravitational force on an object was proportional to its inertial mass. Why was gravitational acceleration independent of the mass or composition of an object? That is to say, why should a cannonball and a feather fall at the same rate?

In addition, Newton believed in the existence of absolute space that did not change or alter. Moreover, the conception of space, in his view, was not dependent on the location of the observer. To prove his contention he performed an experiment in which he hung a bucket full of water from a rope. He rotated the bucket so that the rope was twisted and then let it go. Initially the water surface was flat. As the bucket started spinning around, however, the water gradually picked up the rotation from the bucket. Eventually the water rotated at the same speed as the bucket. At this point the surface of the water was curved into a parabola. Newton maintained that it wasn't the motion of the bucket that changed the surface of the water, because by the time the surface of water was affected the water wasn't moving relative to the bucket anymore. Instead he believed that it was the motion of the water itself that made the difference. Somehow the water was aware of the fact that it was in rotation. This led Newton to conclude that there was an absolute space that decided what did and didn't have a force acting on it. Newton's theory was refuted by Ernst Mach (1838–1916), an Austrian physicist who did most of his major work at the end of the nineteenth century. Mach argued that the water was responding to the mass around it, just as the Earth did. The water, said Mach, was rotating relative to the surrounding mass and not to its own motion. Newton was mistaken to think that there was any absolute space.

In spite of the objections to Newton's theory, no scientist had been able to come up with a more accurate theory of gravity that could address the flaws in his theory. Improbably, the solution would emerge in a burst of unprecedented scientific inspiration due completely to the genius of a clerk working alone in the Swiss patent office.

Albert Einstein was born in Ulm, Germany, on March 14, 1879, and spent his youth in Munich, where his family owned a small shop that manufactured electric machinery. Albert's parents, who were nonobservant Jews, moved from Ulm to Munich when Albert was an infant. The family ran a business manufacturing electrical equipment, but in 1894, after repeated financial failures, the Einsteins left Germany for Milan, Italy. Albert, who was then fifteen years old, stayed in Milan for only a year before going off to finish his secondary education in Switzerland. His academic record was hardly a model of stellar achievement. He frequently cut classes, preferring to spend his time studying physics on his own and playing the violin. (Luckily his talent for the former vastly outweighed his ability with the latter.) Within a year, still without having completed secondary school, Einstein failed an examination that would have allowed him to pursue a course of study leading to a diploma as an electrical engineer at the Swiss Federal Institute of Technology, known as the Zurich Polytechnic. He spent the next year in nearby Aarau at a secondary school, where he enjoyed excellent teachers and first-rate facilities in physics. In 1896 Einstein returned to the Zurich Polytechnic, where he graduated in 1900 as a secondary school teacher of mathematics and physics. His mathematics professor, the German Hermann Minkowski, considered Einstein a "lazy dog" who seldom came to class. His other professors held an even more dubious view of him and refused to recommend him for a teaching position at the University of Zurich.

After two years of struggling to eke out a living as a tutor and substitute teacher, Einstein finally obtained a post as an examiner at the Swiss patent office in Bern. For all the demands that the patent office made on him in the seven years he was employed there, Einstein still managed to find the time to revolutionize human understanding of the universe. In a feverish torrent, beginning in 1905, when he was twenty-six, Einstein produced four seminal papers in as many months.

In March 1905, Einstein overturned the long-held theory that electromagnetic energy—also known as light—consisted of waves (invoked in Newton's conception of gravity) propagating in a hypothetical all-pervasive medium called the luminiferous ether. Einstein argued instead that light could consist of discrete bundles of radiation, now called photons. His theory formed the basis for much of quantum mechanics, and led to the idea that we live in a quantum universe built out of minute, discrete chunks of energy and matter, rather than a flowing, continuous world.

In April and May, he published two more papers; in one he invented a novel method of counting and determining the size of the atoms or molecules in a given space, and in the other he explained the phenomenon of Brownian motion, which made significant predictions about the motion of particles that are randomly distributed in a fluid. The net result of these two contributions provided proof once and for all that atoms actually exist—surprisingly, still a matter of some controversy at that time—and put an end to a millennia-old debate on the fundamental nature of the chemical elements.

Then in June, Einstein completed his special theory of relativity. If Einstein's March paper treated light as particles, special relativity conceived of light as a continuous field of waves. In fact, light behaves like both particles and waves. Light particles are actually packets of energy known as photons; unlike other minute types of matter, such as specks of dust, they have no mass and always move at a constant speed of 186,000 miles per second, when they are in a vacuum. However, when light diffracts, or bends slightly as it passes around a corner, it assumes a wavelike form. Because these wavelike forms consist of changing electric and magnetic fields, they are known as electromagnetic waves. There are two basic postulates to the theory:

- **Postulate 1**: There is no such thing as absolute space or absolute time or an absolute "rest frame" that allows the observer a privileged place in the universe. No matter where any observers are positioned, they can consider themselves to be "at rest." If two observers are moving relative to each other, no experiment can be devised to determine whose observation is "right" or "wrong."

- **Postulate 2**: The speed of light is measured the same by all observers, regardless of the relative speed of the source of light or the location (or speed) of the observer. Einstein based his new idea on a reinterpretation of the classical principle of relativity, which maintained that the laws of physics had to have the same form in any frame of reference. As a second fundamental hypothesis, Einstein assumed that the speed of light—186,000 miles per second—remained constant in all frames of reference. In doing so, he was able to discard the hypothesis of the ether, for it played no role in his conception of the universe. One consequence of this theory is the dilation of time—that is to say that time slows down as one measures its passage on an object moving relative to the observer. The effect is ordinarily very small unless speeds approach the speed of light. The special theory of relativity holds that the speed of light constitutes a "barrier"—nothing can move faster than light, and only packets of energy that do not have any mass whatsoever are capable of moving at the speed of light.

Einstein believed that a good theory is one in which a minimum number of postulates is required to account for the physical evidence. Nowhere is this belief better illustrated in special relativity than in its mathematical description of the relationship of energy and mass which, Einstein contended, are equivalent. This simple statement is described in possibly the most famous (and certainly the best known) equation of all time: $E = mc^2$, where m is the mass of an object, E is its energy content, and c is the velocity of light. In composing this remarkably economical little equation, which unifies the concepts of energy and matter and relates both to the velocity of light, Einstein initially was concerned with mass. This equation accounts for the thermonuclear processes that empower the stars. It also accounts for the explosive power of the atomic bomb.

But for all its extraordinary achievements, special relativity had nothing to say regarding the other known large-scale force in the universe: gravitation. What special relativity does speak to is *inertial mass*, or the resistance objects offer to change in their state of motion—their "clout" or "heft," so to speak. Inertial mass is different from gravitational mass. Gravity acts on objects according to their gravitational mass—what we usually refer to as their "weight." Or to put it another way, inertial mass is what you feel when you slide a suitcase along a polished floor; gravitational mass is what you feel when you lift the suitcase. This would suggest that there ought to be a distinct difference between the two masses. After all, gravitational mass manifests itself only in the presence of gravitational force, while inertial mass is a permanent property of matter. Take the suitcase on a spaceship and it will weigh nothing once the spaceship is in orbit and beyond the gravitational pull of Earth. (Its gravitational mass will measure zero). Yet its inertial mass—its heft—will remain the same. You'll have to work just as hard to wrest it around the cabin. And once the suitcase is in motion, it will have the same momentum as if it were sliding across a floor on Earth. Yet for some reason, the inertial and gravitational masses of any given object are exactly the same—a phenomenon known as equivalence. Say that the suitcase is placed on a scale (on Earth) and it turns out to weigh 30 pounds. That is its gravitational mass. If you then placed the same suitcase on a relatively frictionless surface, attached a spring scale to it, and pulled it until it reached the same rate of acceleration it would have if it were falling (i.e., 16 feet per second, on Earth), the scale would register 30 pounds—a measurement of its inertial mass. Any number of similar experiments have been performed on a variety of objects of differing weights, but no matter how you cut it, the result is invariably the same: there is no difference between inertial mass and gravitational mass.

The equality of gravitational mass and inertial mass had been known for centuries. It explained, for instance, why cannonballs and bocce balls fall at the same velocity in spite of the fact that their weights are so different. Even though a cannonball obviously has a greater gravitational mass (it weighs more), it also has a similarly greater inertial mass, which makes it accelerate more slowly. Or to put it another way, the two masses cancel each other out. But in Newtonian physics, the equivalence principle was regarded as a mere coincidence. Einstein, however, thought that there might be more to it than that.

In his attempt to grasp the significance of equivalency, Einstein realized that if gravity were to be interpreted as a form of acceleration, then that acceleration would have to take place along the undulations of curved space. Why should space be curved? Newton had therorized that every object attracts every other object in direct proportion to its mass. But Einstein maintained that in the neighborhood of very massive objects, such as the Sun, the object's gravitational pull exerts so much strength as it rotates that it "drags" space (and time) along with it, in effect bending or curving the nearby region of space. (This so-called frame-dragging effect was confirmed in 1997 by a group of Italian astronomers observing very dense, rapidly spinning astronomical objects called neutron stars.) An accelerating object doesn't move in three dimensions, it actually moves in four—three of space and one of time. Classical Euclidean geometry offers an accurate description of a two-dimensional world ("plane" geometry) and of a three-dimensional world ("solid" geometry). But if time is taken into account, then things become much more complicated. For years, beginning in 1907, Einstein struggled to puzzle out how gravity acts on curved space. "In my life I have never before labored so hard," he wrote to a friend. "Compared with this problem, the original theory of relativity is child's play."

In keeping with his dictum that the best theories consist of a minimum number of postulates, he sought to find patterns of beauty and simplicity. His search led him to the general theory of relativity. Einstein's initial grasp of a solution to the question of gravity came in a flash of insight one day in 1907, while at work in the patent office, in what he later called "the happiest thought of my life." The memory of the moment remained vivid in his mind decades later.

The flash of insight came about by means of a thought experiment. Einstein began with the image of a spaceship sitting still on the Earth. If you were in the spaceship, and you threw a ball across the ship, it would fall to the floor because gravity would attract the ball toward the Earth, which is also in the direction of the floor. Now suppose that the spaceship was far out in space, beyond the gravitational pull of any large bodies like planets or stars. If you threw the ball across the spaceship, it would travel in a straight line because there would be no force acting on the ball to cause it to deviate. If, however, the spaceship was accelerating (it's not just moving, but gaining in speed) and you threw the ball across the ship, the ball would still move in a straight line, since there is no force acting on the ball. However, as the ball travels across the spaceship, it would appear as if it were getting closer to the floor, because the acceleration of the spaceship is causing the floor to move up toward the ball! If you were a passenger on the spaceship, you would also be moving up toward the ball. In fact, the motion of the ball would look no different to you if you'd thrown the ball on an accelerating spaceship than if you'd thrown it on a spaceship sitting still on the Earth. Einstein's great revelation was to recognize that the acceleration of the spaceship had the same value as the acceleration due to gravity on Earth. Effectively, you cannot tell the difference between being at rest on Earth and subject to Earth's gravity, and being out in space in an accelerating ship. Or to put it another way, you cannot tell the difference between observations of objects made while in an accelerating ship and observations of objects made while in the presence of gravity. What Einstein had come up with was the equivalence principle—the puzzling equality of gravitational mass and inertial mass—in another guise.

Why should this insight have stirred such elation in Einstein? Try another thought experiment: Imagine that you awaken to find yourself floating, weightless, in a sealed, windowless elevator car. On the wall you find a set of instructions informing you that you are in one of two identical elevator cars—one adrift in deep space where it is free of almost all gravitational influence, and the other located on Earth and now rapidly plummeting to its doom. The only way you can be rescued is to *prove* (not guess) whether your elevator car is the one where gravity does not operate or the one in which gravity is operating all too well. What Einstein realized was that it is impossible to tell the difference, neither through your senses nor by conducting experiments. Just because you are weightless doesn't necessarily mean that you are free from gravity; it might mean only that you might be in free fall—gravity has no effect on you even if you happen to be on Earth. That is to say that if both you and the elevator you are in are in free fall, you are as "weightless" as an astronaut in a spaceship far out in space.

Taken one step further, Einstein postulated that the gravitational field has only a *relative* existence. The same ambiguity applies in the opposite situation: Suppose that when you awaken you find yourself in the elevator car, at your normal weight. Now the instructions on the wall explain that you are either in an elevator car stopped on the ground floor of an office building on Earth, or else adrift in zero-gravity space, in an elevator attached by a cable to a spaceship that is accelerating, pressing you to the floor with a force equal to that of Earth's gravity—or 1 g. Again you would be unable to prove that you are in one elevator or the other.

By the time Einstein had hit on his gravitational theory, he had begun to gain enough professional recognition to allow him to quit the patent office and accept a series of teaching positions. Eventually he would be appointed to a full professorship in pure research at the University of Berlin. But the puzzle of gravity continued to gnaw at him. From his thought experiments Einstein reasoned that gravity itself might be regarded as a form of acceleration. But acceleration through what reference frame exactly? It couldn't be ordinary three-dimensional space; the passengers descending in an elevator in an office building are not flying through space relative to the Earth, after all.

In 1912, Einstein began a new phase of his gravitational research, with the help of his mathematician friend Marcel Grossmann. For his new work Einstein began to rely on something called tensor calculus, devised by Italian mathematicians Tullio Levi-Civita (1873–1941) and Gregorio Ricci-Curbastro (1853–1925). Tensor calculus greatly facilitated calculations in four-dimensional space-time, a notion that Einstein had obtained from Hermann Minkowski's 1907 mathematical elaboration of his own special theory of relativity. (It was Minkowski who had called Einstein a "lazy dog.")

Simply stated, space-time refers to three-dimensional space plus time, taken together to locate something. That is, all events occur somewhere in space—plotted as points (x, y, z)—and time (t), so they have four "coordinates" that locate them: x, y, z, t. The work would culminate in Einstein's general theory of relativity.

Gravity, Einstein concluded, is not a force acting over a distance like magnetism, as Newton believed, but rather is caused by dips and curves in the fabric of space and time made by the objects residing in it. In other words, in Einstein's view, gravity is not really a property of matter, which Newton believed. It is, instead, a manifestation of the effect of matter on the space-time around it. The dilemma confronting Einstein was this: An object can only exert a force on another object when the two are in contact, and it does so only when the second object stands between the first object and its "natural" state (at rest on the ground for heavy objects, up in the air for light stuff like steam, flame, etc.). A rock in free fall on Earth, for example, is traveling toward the ground (its natural state). If you happen to get in the way, you will be preventing the rock from reaching its natural state (much to your peril), at least temporarily. But how is this force "transmitted" across the empty space in between? We can give up spooky action-at-a-distance by requiring a source of gravity, like the Earth, to emit a "field" (you can think of it as a fluid without any mass) that permeates all of space. Any other matter coming in contact with the field will feel a force exerted by it. So in effect, nothing acts at a distance. But what physically is this field? It isn't tangible like matter, so how can we be certain that it exists if the only evidence of it is its gravitational effects? It almost seems fake, artificial, very much like the infamous ether.

Einstein's solution to the problem married the Aristotelian concept of a natural state with the Newtonian concept of inertia. Inertia refers to the tendency of objects traveling in straight lines to continue doing so unless some force interferes and deflects it. But what is this force? Einstein decided to dispense with it altogether. Instead he approached the problem from another direction, concentrating his focus on the nature of a "straight" line. How, he asked, does the straight line change when a mass is present? Masses like planets and stars, he proposed, bend or warp space. The reason that objects traveling near large masses move in curves instead of in a straight line is that the mass has caused space to bend! Think of a racing car traveling on a banked curving track; the curvature of the track causes the car to travel around the corner rather than hurtle straight ahead. Because the presence of mass curves space-time, "straight lines" are no longer straight in the way we ordinarily think of them. Imagine two people back to back on the North Pole starting to walk out in what they think are straight lines. If the Earth were flat, the people—like the lines—would never meet. But the Earth is not flat and we can predict that the two will eventually meet when they reach the South Pole.

Another problem arose: Mathematically, how could the curvature of space be represented? One way, Einstein thought, was to make the curvature less curved by magnifying a small portion of it. We can continue our focusing into smaller and smaller sections of the curved line. Eventually we'll be looking at such a small portion of the curved line that it will appear straight to us, much the way the Earth seems flat at close range even though we know that it is spherical. This analogy led Einstein to conclude that the gravitational field is in fact a statement about the geometry of space-time itself. The presence of massive objects like the Sun causes space-time to curve in much the way that the surface of the Earth is curved. In curved space the rules of Euclidean geometry are changed. Parallel lines can meet, and the sum of the angles in a triangle can be more, or less, than 180 degrees, depending on how space is curved.

The curvature of space-time is much harder to visualize, though, than the curved surface of a balloon or Earth. The Earth is a two-dimensional curved surface occupying a three-dimensional volume. While it's far harder to imagine a curved three- or four-dimensional volume, it turns out that the mathematics is very similar and it is possible to carry over the same mathematical principles from curved two-dimensional surfaces to higher dimensions. In the absence of mass (empty space instead of space occupied by planets or stars, for example), the geometry of space-time is flat, in the way the tabletop is a flat two-dimensional surface. When, however, a mass is present, the space-time curves to act like the surface of Earth or the surface of water in the rotating bucket—except that the curved "surface" is in four dimensions (three dimensions of space and one of time) rather than two.

If the gravitational force isn't very intense—if it's the Earth's gravitational force and not that of a star, for example—then Einstein's theory of gravity no longer needs to say anything about the curvature of space-time and becomes perfectly compatible with Newtonian gravitational theory. This is indeed a happy development, since Newtonian gravitational theory offers the best explanation we have for the motions of planets. (Not to mention that it is the most *widespread* explanation.) Where Newtonian gravity failed, as in explaining the rotation of Mercury, Einsteinian gravity rushes to the rescue.

Why is Mercury an exceptional case? Well, because Mercury is the planet in our solar system that is closest to the Sun, its gravitational field is the most intense, causing a distortion in the orbit that Newton's theory did not take into account. And there the corrections made by Einstein's gravity match perfectly with observations. The previously unexplained fast precession of Mercury is exactly what is to be expected once corrections due to Einstein's theory of gravity are applied. It also explains why objects fall independently of their mass: they all follow the same straightest possible line in curved space-time—the path of least resistance.

In Einstein's model, a star like the Sun would put a dent in space-time. In other words, it would cause the space around itself to curve. This curvature would affect the path of an asteroid as it approached the star. The asteroid might be deflected by the curvature, or, if it traveled close enough to the star, the curvature could capture it, causing it to orbit the star. Now we return to the notion of a "field." The asteroid would not be *pulled* by the star's *gravity*. Instead it would just follow the path of least resistance—what it "feels" is a straight line. Imagine that you're standing halfway down the steep slope of a crater. If you want to get to the halfway point of the slope on the other side of the crater, you have two choices: You could walk down the slope to the bottom of the crater, then up the other side until you got halfway up. But why expend the effort? It would be much easier simply to stay on a ledge at the same level (not moving up or down the crater) and walk around to the other side. The walk would be longer, to be sure, but you would be exerting less energy. You'd be taking the path of least resistance. The same thing holds true with an asteroid or planet orbiting a star.

According to Einstein, massive rotating objects wrap space and time around themselves and drag them along as they spin, much like a bowling ball turning in a thick liquidlike molasses. If mass wraps space around itself, then, said Einstein, the result would be a four-dimensionally spherical cosmos, in which any observer in the universe would see galaxies stretching deep into space in every direction. But the observer would be wrong to conclude that there is no limit to space. In spite of what our eyes tell us, the amount of space in a closed universe is nonetheless finite. Adventurers with time to spare might be able to visit every galaxy, yet would never reach an edge of space. Just as the surface of the Earth is finite but unbounded in two dimensions (we can wander wherever we like, and will not fall off the edge of the Earth), so a closed four-dimensional universe is unbounded to us, who only occupy it in three dimensions. But according to Einstein, the observer would be under a misapprehension because the universe actually turns out to be bounded in four dimensions.

How could Einstein's theory of gravity possibly be tested? Recall Einstein's thought experiment with a ball in an accelerating spaceship. The ball appears to be moving toward the floor, or conversely, the floor appears to be moving upward toward the ball. By the same token, Einstein realized that if you were to shine a light across the accelerating spaceship, the floor would still move up toward the light, and it would appear that the light was bending toward the floor just as the ball does. Einstein went so far as to propose that light, too, even though it lacks mass, should respond to gravity as well. If Einstein's theory was correct, then it should be possible to prove that gravity should affect the path of light. By 1911, Einstein was able to make preliminary predictions about how a ray of light from a distant star, passing near the Sun, would appear to be attracted, or bent slightly, in the direction of the Sun's mass. After all, light must pass through space, and if space is being bent, light must bend accordingly.

After several false starts, Einstein published the definitive form of the general theory of relativity in late 1915. Others in the scientific community might have been skeptical about the validity of the theory, but Einstein was undaunted. He had no doubt about the correctness of his system. "The sense of the thing is too evident," he said. In 1919, when a student asked Einstein what would happen if the experimental measurements didn't match his theory, Einstein replied, "I would have felt sorry for the dear Lord, because the theory is correct."

If Einstein was right, it should be possible to observe the effect of the Sun's mass on the light coming from the stars. But because the stars are indiscernible when the Sun is out, scientists had to wait for the next total solar eclipse, which would occur on May 29, 1919, when the Sun would stand against the bright stars of the Hyades cluster. To conduct the historic test, the English astronomer Arthur Stanley Eddington led an expedition to a cocoa plantation on Principe Island off West Equatorial Africa. His objective was to see whether curvature of space in the region of the Sun would manifest itself in the apparent positions of the stars in the briefly darkened sky. Tension heightened when rain clouds moved in, threatening the success of the experiment. But only moments after the Moon's shadow came speeding across the landscape and totality began, a hole opened up in the clouds, triggering the camera shutters.

The results of Eddington's expedition, and of a second eclipse observation, conducted at Sobral, Brazil, on the same day, were presented at a meeting of the Royal Society on November 6, 1919. The light rays emitted by the stars were found to have bent to just the degree predicted in Einstein's theory. Einstein had never had any doubts, of course. Years later, in a conversation about the German physicist Max Planck (1858–1947), the father of quantum physics, Einstein remarked, "He was one of the finest people I have ever known and one of my best friends; but, you know, he didn't really understand physics." When asked what he meant by this, Einstein explained: "During the eclipse of 1919,

Planck stayed up all night to see if it would confirm the bending of light by the gravitational field of the sun. If he had really understood the way the general theory of relativity explains the equivalence of inertial and gravitational mass, he would have gone to bed the way I did."

The experimental confirmation of his theory of gravitation made Einstein a celebrity, lionized by the popular press. Several tests of the theory have been conducted subsequently to Eddington's observations. As recently as April 1998, British astronomers working with NASA's Hubble Space Telescope captured light from a distant galaxy warped by gravity into a perfect "Einstein ring"—just as Einstein had predicted. But the theory continues to be challenged by other observational tests. And there are still predictions of Einstein's theory that have yet to be borne out, including his postulation of gravitational waves. So far these waves have proven extremely difficult to observe and require sensitivity that is yet to be achieved with current technology.

Whether or not gravitational waves can be found, there is no question that general relativity has become one of the most beloved scientific theories in history, and not just because of its success, but also because of its beauty and richness. This could be one of its drawbacks, too, though. "We love it—perhaps too much," remarked physicist and writer Robert Park. "We revere it and that's a mistake. Because when you revere something you don't move on."

CHAPTER 6

The Forgotten Inventor

Philo Farnsworth and the Development of Television

Television is such an ubiquitous presence in our lives and so unrivaled in its influence that it is somewhat surprising that so few people know who invented it. Edison is associated with the invention of the lightbulb, Marconi with the invention of radio, and Bell with the invention of the telephone, but the name Philo Farnsworth is unlikely to ring any bells at all. Farnsworth's story underscores two crucial verities in the history of innovation. The first is that a brilliant idea can germinate in the most unexpected places, and the second is that if you're going to be an inventor it helps to also have a head for business.

Certainly few inventors since have come from humbler backgrounds. Farnsworth was born in a log cabin and rode to high school on horseback; it wasn't until he was 11 years old that he had his first encounter with electricity. Nonetheless, he seemed to have known his destiny even as a child. At the age of six he declared his intention to become an inventor like his heroes Bell and Edison.

In some sense the story of television began in Rigby, Idaho, in the spring of 1919, when the Farnsworth family moved from

Utah, where Philo was born, to an Idaho farm. When their covered wagons reached the crest of the hill, young Philo gazed down into the valley where his future home was situated and remarked on the one thing that had escaped every other member of his family. The farmhouse, barn, and outbuildings were all connected by wires. "This place has electricity!" he cried out excitedly.

Although up until then Farnsworth had only known about electricity from books, he seemed to have an intuitive understanding of this invisible force that he immediately set about putting to use. Cobbling together spare parts lying around the farm, Philo began building electric motors for the washing machine and farm equipment—constructing over a dozen devices in all. As a teenager, he worked part-time repairing radios and thought constantly about the properties of the electron. He taught himself physics, studying Einstein's theories and reading borrowed science books and magazines late into the night. One winter night Farnsworth was leafing through a magazine when he stumbled on a speculative article titled "Pictures That Could Fly Through the Air." The writer envisioned an electronic device that would represent a hybrid of radio and movies capable of simultaneously projecting both image and sound into houses around the world. Once the idea of flying pictures took root in his mind, Farnsworth began to read everything he could lay his hands on about the subject. He soon discovered from his reading that several inventors had achieved limited success with a mechanical television system, but he believed—correctly, as it turned out—that these rudimentary systems wouldn't work fast enough to capture and reassemble anything but shadows and flickers.

Nearly fifty years earlier, a system for transmitting images and sound was proposed by inventor George Carey of Boston. His vision called for each element or piece of the picture to be simultaneously carried over separate circuits. In 1880, W. E.

Sawyer in the United States and Maurice Leblanc in France came up with the operating principle for all forms of televised transmission. Instead of relying on separate circuits, they contended, each element in the final picture could be rapidly scanned, line by line and frame by frame. Because of the limitations of human eyesight, the resulting image would appear as a coherent whole rather than as a succession of black-and-white dots. Now that the theoretical possibility of using only a single wire or channel for transmission was established, the next question for engineers and scientists was how to accomplish it in fact.

One promising avenue was suggested by the discovery of the photoconductive properties of the element selenium in 1873. The finding that the rate of electrical conduction varied with the amount of light paved the way for the next major proposal for a practical television. In 1884 a German inventor named Paul Nipkow (1860–1940) received a patent for a system based on a rotating disk with a spiral-shaped aperture. What distinguished Nipkow's device was that it was able to scan images simply and effectively at both the sending and receiving ends.

In Nipkow's system, the image to be televised is focused on a rotating disk with square apertures arranged in the form of a spiral. As the disk rotates, the outermost aperture traces out a line across the top of the image. As it passes through the aperture, the light varies in direct proportion to the light and shade of the line of the image. These variations are known as brightness values. Once the outermost aperture has passed over the image, the next inner aperture repeats the process and traces out a second line of the image, immediately below the first. As the disk continues to rotate, successive lines are traced out, one beneath the other, until the whole image has been traced. The more apertures available on the disk, the greater the detail in the image that can be defined.

In the next step of the process, the light passing through the apertures enters a photoelectric cell that translates the sequence of brightness values into a corresponding sequence of electric values. These impulses are then transmitted over a single circuit to the receiver, usually a screen. Then the process has to be reversed. The electrical impulses have to be turned back into light again. In early systems, this was accomplished by using a lamp capable of reproducing the sequence of brightness values. The light from the lamp is projected onto the surface of a disk resembling the transmitter. This disk repeats the scanning process of the first disk, only in reverse. While it rotates in precise synchronism, the disk scans each line of the image, turning the electrical impulses into brightness values until the lines are reassembled in their proper order and the original image is reproduced in its entirety. Of course, the rotation has to happen at a speed fast enough so that the eye perceives the image as a whole rather than as a series of moving points. Whatever the limitations of Nipkow's device, the basic principle of synchronizing the scanning speed of the camera and the receiver remains essential to all television systems in use today.

The major stumbling block in practical terms to Nipkow's television was the means of transmission. Selenium responded to changes in light too slowly to make it a very suitable photoconductor. That led researchers to look for another photosensitive material that could be employed in its place. In 1913 German scientists created a potassium hydride–coated cell, which offered heightened sensitivity to light and had the added advantage of being able to follow rapid changes of light as well. The introduction of this cell made a practical working television system possible for the first time.

At the same time as work was under way to develop a more efficient transmitter, other researchers were concentrating their efforts on the receiver. In 1897 Karl Ferdinand Braun, a German inventor, pioneered an early receiver based on a cathode-ray tube and a fluorescent screen. (A cathode-ray tube, or CRT, is an electron tube that converts electrical signals into a pattern on a screen and forms the basis of the television receiver.) The beam of electrons would pass through the tube and strike the screen, producing visible light in response. In 1907 a Russian scientist, Boris Rosing proposed the use of Braun's receiver and then improved upon it by introducing a mirror-drum scanner that operated at the transmitter end. With this method he actually succeeded in transmitting and reproducing some crude geometrical patterns.

The next breakthrough came as a result of work by Scottish electrical engineer A. A. Campbell Swinton. Between 1908 and 1911, Swinton proposed a method that, in all essentials, is the basis of modern television. His contribution was to use cathode-ray tubes at both the camera and receiver ends. In his scheme the image of the scene to be transmitted was focused onto a mosaic screen of photoelectric elements; then a cathoderay beam, directed on the back of the screen, would trace out a line-by-line scanning sequence. Swinton's ideas, however, were too advanced for practical application.

Six years later an American engineer named Daniel McFarlan Moore created the first neon gas–discharge lamp, which made it possible to vary the light intensity at the receiver end. By varying the electrical input to the neon lamp, Moore found, he could produce modulated light.

It was at this point that Philo Farnsworth entered the picture. Entranced by the possibility of producing "flying pictures," Farnsworth was determined to learn everything he could about the subject. It soon became apparent to him, though, that there was an inherent problem in trying to convert light into electricity using whirling disks and mirrors, for the simple reason that they could never whirl fast enough to transmit a coherent image. What was needed was a means of transmission that could work at the speed of light itself.

The solution to the problem that would change the twentieth century arrived in a burst of illumination—appropriately enough—while Philo was chained to a horse-drawn harrow (a harvesting machine), endlessly crisscrossing the fields of his family farm, row by row, harvesting potatoes. To relieve the boredom of the dreary task, he began to dream about television. What if light could be trapped in a kind of empty jar and then transmitted? But how would the transmission be accomplished? Ironically, the very nature of the job he was engaged in gave him the inspiration he needed to reach the solution. The image could be scanned on a magnetically deflected beam of electrons in much the same way that the harrow harvested potatoes, working line by line, duplicating the way the eye works when you read a book. Though the essence of Farnsworth's idea is extraordinarily simple, it took a 14-year-old boy to come up with it. The principle that Farnsworth conceived of that day in the potato fields remains today the basis for modern television.

It was one thing to come up with a bright idea, but now Farnsworth has confronted by the problem of testing it to see whether it would work in practice. To appreciate the difficulties the boy faced, you have to keep in mind that some of the most brilliant scientists, subsidized by some of the biggest electrical companies of the world, were hard at work on the same problem that this farm boy had just solved. Moreover, he had little education, no connections, and no money. His father was farsighted enough to realize that his son was onto something, but he advised him not to discuss his idea with anyone. If the idea was as valuable as the elder Farnsworth suspected, it was likely to be pirated. But while secrecy was prudent, Philo still had to find out whether his idea could work. What he needed was someone in whom he could confide, who would also understand what he was trying to get at.

That was when he thought of his high school chemistry teacher, Justin Tolman. Farnsworth had already developed a close relationship with Tolman. The teacher was so impressed by his pupil's scientific precocity and enthusiasm that he had agreed to give him special instruction in private and allowed him to audit a senior course. But as much as he appreciated Farnsworth's aptitude, he was hardly prepared for what was waiting for him late one afternoon in March 1922. Walking into his classroom, Tolman was arrested by the sight of his young prodigy covering the blackboard with complicated electrical diagrams and equations.

Tolman couldn't make heads or tails of what the boy was writing. Farnsworth explained that he had come up with an idea that he was anxious to confide in his teacher since he was convinced no one else would understand it. He went on to say that he had figured out a way of making an electronic television. Tolman was as mystified as before: he'd never heard of television.

He was soon to find out.

The two spent the next several weeks developing and elaborating on Farnsworth's idea. By the time the semester ended, they were both convinced that the idea could work. But the idea was still confined to the blackboard and notepads. When it would be possible to actually test it was a question that neither teacher nor pupil could answer.

Forced to move by hard times, Philo's family relocated to more fertile soil, near Provo, Utah. They shared a house with another family, the Gardners. Their neighbors had eight children—two sons and six daughters, something that wouldn't be worth noting if it weren't for the fact that two of the Gardner children were to play a significant role in Farnsworth's life and career.

Demonstrating the same determination that he had shown in high school, Farnsworth succeeded in gaining admission as a special freshman to Brigham Young University. For the first time he had the resources of a major university to draw on for his research, which was increasingly focused on cathode-ray tubes and vacuum tubes. But without the financial wherewithal, he was no closer to constructing a prototype of the device that he could see so vividly in his mind's eye.

Tragedy interrupted his studies in the winter of 1923, with the death of his father. Farnsworth was now saddled with the responsibility for taking care of the family. (He even began to call himself "Phil" because he thought it made him sound more adult.) And with no money coming in, he was forced to leave Brigham Young to find work. His dream of inventing an electron-based television seemed even more remote than before. There was one bright spot in his life: Farnsworth had begun to spend more and more time with the prettiest of the six Gardner daughters, Elma, whom everyone called Pem. It soon became obvious that they were meant for each other, and Farnsworth proposed to Pem on her birthday in February 1926. With so much uncertainty in their lives, though, they postponed setting a date for the wedding.

In the meantime, Farnsworth and his future brother-in-law, Cliff Gardner, had subscribed to a correspondence course in radio maintenance. Once they were confident that they had the necessary expertise, they established their own radio installation and repair business in the spring of 1926. Farnsworth's first foray into business didn't go well, though—a harbinger of things to come. Out of desperation, Farnsworth told Gardner that he was thinking about writing up his ideas for designing a television and submitting it to *Popular Science* magazine. If everything worked out right, he said, maybe he could make as much as a hundred dollars. Knowing what he did about his friend's ideas, Gardner was appalled. He convinced Farnsworth that publishing would be a mistake he would later regret. Farnsworth then signed up with the University of Utah placement service in hope that it might help find him work. As it turned out, his choice proved serendipitous.

In the spring of 1926, George Everson, a professional fundraiser with connections to some of the West Coast's richest financial circles, was in Salt Lake City recruiting college students to carry out a community survey. One of his most enthusiastic applicants was the 19-year-old Philo T. Farnsworth. As the survey was winding down, Everson asked Farnsworth if he was planning to go back to school. Farnsworth told him that he couldn't afford to, and besides, he was more preoccupied by his attempt to find a way to finance a new invention of his. "I've been thinking about it for about five years, though, and I'm quite sure it would work," he said. "Unfortunately, the only way I can prove it is by doing it myself; but I don't have any money."

When Farnsworth explained his idea, Everson and his partner, Les Gorrell, had no idea what he was talking about. But they had no trouble discerning his passion. As Farnsworth began to elaborate on his conception, his whole personality seemed to undergo a dramatic transformation. Suddenly he was no longer like "an office clerk too closely confined to his work," in Everson's words.

All the same Everson remained skeptical, since he couldn't believe that either GE or Bell Labs hadn't already succeeded in accomplishing what Farnsworth proposed. Farnsworth was prepared to counter any argument. He proceeded to summarize the progress that had been made to date in realizing a viable television system. Mechanical systems, he contended, lacked the necessary sensitivity. The weaknesses of these systems was underscored by every attempt to increase the number of lines in hope of improving the definition of the pictures. If television was to work, Farnsworth said, it would have to show pictures that had both good quality and definition on a screen of reasonable size. But to meet these criteria, as Swinton and others had pointed out, the image would need to be defined in enough detail to generate at least 100,000 and preferably 200,000 elements.

Since the number of elements is approximately equal to the square of the number of lines, it was obvious that any system using 30 or even 100 lines would be inadequate. This was one of Farnsworth's principal objections to the results of an experimental demonstration conducted earlier that year by a Scottish engineer named John Logie Baird. Baird was the first to have shown that it was possible to electrically transmit moving pictures in halftones. These pictures were formed of only 30 lines, repeating approximately 10 times per second. The results were predictably crude: the receiver screen was dim and only a few inches high, and the picture flickered badly.

It was true that mechanical systems could be made to operate on 200 and more lines, though with great difficulty. Without electricity, sequential scanning depended on mechanisms such as mirror drums or disks with lenses. To Farnsworth's way of thinking, engineers who were following this path were all "barking up the wrong tree."

Increasingly intrigued, Everson finally asked Farnsworth how much it might cost to build a model of the machine. Farnsworth picked a figure of $5,000 out of thin air. "Well," Everson said, "Your guess is as good as any. I surely have no idea what is involved. But I have about $6,000 in a special account in San Francisco. I've been saving it with the idea that I'd take a long shot on something and maybe make a killing." Acknowledging that it was "about as wild a gamble as I can imagine," he agreed to stake Farnsworth to the whole $6,000. "If we win," he said, "it will be fine, but if we lose, I won't squawk."

In view of the fact that he had conceived of the idea in the first place, Farnsworth insisted on retaining nominal control of the association formed between him, Everson, and Everson's partner Gorrell. He was also awarded half the equity in the company. After some debate, they agreed to set up operations in San Francisco. The unexpected turn in Farnsworth's fortunes also had another benefit in that he and Pem were now in a position to go ahead and make plans for their wedding.

Pem's brother, Cliff Gardner, was folding cardboard into boxes on an assembly line in Oregon when a telegram from Phil and Pem arrived. The message said only that Farnsworth had found a source of financial support and that he should come at once to San Francisco.

As soon as the newlyweds and Gardner were reunited, they went to inspect 202 Green Street, the location of the loft that the bankers had provided for Farnsworth's new laboratory. No sooner had Farnsworth set foot in the place than he became immediately convinced that he had arrived at the birthplace of television.

Before anything else, though, Farnsworth wanted to make certain that his idea was protected. After he had finalized the plans for his television system and drawn detailed diagrams, he filed for his first patent, on January 7, 1927. No patents could be officially granted, however, until the device had been proven to work, or "reduced to practice," in the parlance of the U.S. Patent Office. And a great deal of research and testing still remained to be done before that could occur.

After a few months, the two men felt confident enough to undertake the building of the world's first electronic television camera tube, which Farnsworth christened the Image Dissector. The device earned its name because it was capable of transmitting an image by dissecting it into individual elements and then converting the elements—one line at a time—into a pulsating electrical current. To construct the photoelectric surfaces of the Image Dissector, Farnsworth and Gardner made use of cesium, a soft, metallic, radioactive element. Cesium was chosen because of its property of emitting electrons when exposed to light. The problem for the two was to get hold of a sufficient quantity of cesium (it ranks forty-sixth in abundance among the natural elements). The solution was to purchase as many cases of radio tubes as they could. Radio tubes were a good source because they used small cesium pellets to absorb any gases that remained after all the air had been pumped out of them.

Farnsworth also realized that a different kind of cathode-ray tube would be required if his system was to work properly. So he invented a cold cathode-ray tube, which, unlike most cathode-ray tubes, could convert electrical signals into a pattern on the screen without being heated.

The next step was to put together the receiving end of the system. To start with, Farnsworth used a standard Erlenmeyer flask, similar to the flat-bottomed ones he had used in his high school chemistry class, and appropriated it for the first "picture tube," which he dubbed the Image Oscillite.

Altogether, Farnsworth and Gardner spent a year laying the foundation for television. Late in the summer of 1927, they rigged together a rudimentary apparatus and began testing it to see if the system could send an image from the camera to the receiver. The first few tests were disappointing. The receiver would light up when the current flowed through the cathode-ray tube, but there was nothing to see but a blur of electronic interference. Rather than let himself get too discouraged, Farnsworth analyzed the results from each test and then set about refining parts of the system to address the problem.

On September 7, 1927, the system was ready to be tested again. This time Farnsworth was so confident that he invited Everson and Pem to the lab to witness what he hoped would be his inaugural "transmission."

To conduct the test, Farnsworth chose the simplest of images. He painted a glass slide black with a straight white line down its center. There was a compelling logic to his choice of image for this test. If the spectators could tell by looking at the receiver whether the line was vertical or horizontal, then they could be certain that they were seeing a transmitted image. In another room, Gardner dropped the glass slide between the Image Dissector and a hot, bright, carbon-arc lamp while Farnsworth, Pem, and Everson waited expectantly in the adjoining room. The screen flickered and gyrated for a moment, but then resolved itself into the image of a straight line shimmering boldly against an eerie blue electronic hue on the bottom of Farnsworth's tubes. Then, as Gardner rotated the slide 90 degrees, the three spectators saw it move—which is to say that they were seeing the first exclusively electronic television picture ever transmitted from one place to another. Farnsworth's reaction was characteristically tempered. "There you are," he said, "electronic television."

Farnsworth's predecessors had struggled to perfect a mechanical system, using large, spinning perforated disks to scatter light into a glass screen in hope that somehow the alternating patterns of light and dark would result in an image similar in quality to images that moviegoing audiences were used to. But for all their efforts over ten years, these engineers had only succeeded in producing blurry, inchoate images, about the size of a postage stamp. What Farnsworth had accomplished was to build a device that was capable of transmitting images in far more detail by transmitting them one line at a time on a magnetically deflected beam of electrons in the same way that you scan each line of a page in a book. This principle still remains the basis of modern analog television.

Later that evening, Farnsworth documented the arrival of true television with characteristic understatement, noting in his laboratory journal, "The received line picture was evident this time."

But in a telegram to Les Gorrell in Los Angeles, George Everson put it much more succinctly—and boldly: THE DAMNED THING WORKS!

It took a year for Farnsworth's breakthrough to reach the public. SF MAN'S INVENTION TO REVOLUTIONIZE TELEVISION, proclaimed a headline in the *San Francisco Chronicle* of September 3, 1928. A new word had just been added to the nation's vocabulary. The accompanying article described the Image Dissector as being the size of "an ordinary quart jar that a housewife uses to preserve fruit" and the result produced by Farnsworth's new system as a "queer looking line image in a bluish light which smudges and blurs frequently." However, the significance of the invention didn't escape the reporter, who noted, "but the basic principle is achieved and perfection is now a matter of engineering."

It didn't take long for the electronics industry, based in the East, to learn of Farnsworth's innovation. While patents were pending, however, vital details relating to Farnsworth's work remained a closely guarded secret. In any event, many inventors who were developing television systems of their own, still relying on discredited mechanical methods, failed to appreciate the significance of Farnsworth's invention. One person who did take a great interest in the work, though, was the recently appointed vice president and general manager of the vast Radio Corporation of America, David Sarnoff. He was intent on finding out exactly what was happening at 202 Green Street.

A temperamental and ambitious Russian immigrant, Sarnoff reputedly got his first taste of the power of modern communications the night he reported the sinking of the *Titanic* to the world from his post on the wireless for American Marconi, a legend that has since been debunked. With RCA's patents on radio close to expiring, Sarnoff was eager to get his hands on this new type of radio-with-a-picture before any competitors could move in. His strategy was to secure patents on television long enough to stall new developments in the invention so he could gain the maximum return on the patents he still held on radio. Sarnoff had some reason to worry. Every time there was a flurry of publicity about television, the sales of radio sets went down because consumers preferred to save their money and wait for the new device. Sarnoff feared that the new television industry would soon make the radio industry obsolete. His solution was to start a television company of his own.

In 1930 Sarnoff contracted the services of Vladimir K. Zworykin, a Russian-born research engineer who had some experience in television. Zworykin had originally filed for a patent in 1923 for a camera tube called an iconoscope. Although the Russian had achieved some significant results with a receiver similar to Farnsworth's in 1929, two years after Farnsworth, he was unable to duplicate the latter's success with a suitable electronic camera device. So he ended up resorting to the use of spinning wheels on the input end even into the early 1930s. His work on the picture tube was retarded by the same limitations that had bedeviled previous researchers. His system could not produce more than 40 or 50 lines per frame because the receiver was incapable of displaying any more detail than was sent by the transmitter.

Reassembling an image on a photo cathode in an evacuated bottle—converting electricity back into light—was the easy side of the equation. The hard part was converting values of light into values of electricity, and that was something Zworykin couldn't do. The true brilliance of Farnsworth's work was his ability to solve the conversion problem.

Without telling Farnsworth the identity of his new employer, Zworykin called on the inventor in San Francisco, introducing himself as a fellow researcher interested in television. According to all accounts, Farnsworth was a welcoming host, allowing Zworykin the run of his laboratory for three full days, during which time his visitor had ample opportunity to acquaint himself with many of the most confidential aspects of Farnsworth's work. Shown the Image Dissector, Zworykin was overheard to say, "This is a beautiful instrument. I wish I'd invented it." It was a remark that would come back to haunt both men.

Returning to RCA's Camden, New Jersey, labs, Zworykin began to try to reverse-engineer Farnsworth's invention. Sarnoff was so confident that he would succeed that he provided his employee with a $100,000 budget, far more than Farnsworth had been able to raise. Sarnoff gave Zworykin a year to invent a television of his own. But the deadline passed and Zworykin still had not produced one.

To head off any threat from Farnsworth, Sarnoff tried to buy him out. Although Farnsworth had applied for two key patents of his own, his financial backers were pressuring him to sell the whole company—and not just a license—to Westinghouse. The stock market crash of 1929 had made them eager to get as much money as they could out of their investment. So Sarnoff assumed that it would be relatively simple to buy the inventor out. He quietly offered Everson what was then a staggering sum—about $100,000—for the company. He insisted that the deal include the services of Farnsworth. Everson rebuffed him, though, knowing that Farnsworth would be certain to refuse. And even though he was anxious to bail out of the enterprise with as much money as he could, Everson also realized that the new invention was worth considerably more than Sarnoff was proposing. "Well, then," Sarnoff said, "there's nothing here we'll need."

As both Farnsworth and Everson realized, it was no longer possible to go it alone. They found a partner in Philco, which had a fair share of the radio business during the 1920s, for which it paid usual patent royalties to RCA. But Philco was hardly a major player in the electronics industry compared to the so-called Radio Trust, in which large companies like RCA, AT&T, and General Electric all pooled their patents. The Green Street team then relocated to Philadelphia to continue perfecting the invention, this time with corporate money behind them. Gradually the images that Farnsworth produced began to acquire the vividness and clarity that he and his colleagues had been struggling to achieve ever since they started.

By 1933 Farnsworth had obtained an experimental license from the FCC to conduct over-the-air television transmissions. He set up a prototype receiver in his home, where he could count on his young son, Philo III, to provide a faithful audience. His programming mainly consisted of a Mickey Mouse cartoon, "Steamboat Willy," which was broadcast from the Philco laboratory several miles away.

Meanwhile, Sarnoff continued his own efforts to try to monopolize the nascent television industry. In 1932, employing an improved cathode-ray tube for the receiver, RCA demonstrated an electronic television that produced an initial image of 120 lines, twice the number that Farnsworth's rudimentary device was capable of. The innovative work that made RCA's system possible was credited to Zworykin, who'd produced his new camera tube—called the Iconoscope—three years after his visit to Farnsworth's lab in San Francisco. While Zworykin had made a significant contribution to the camera tube, much of his work was influenced by research carried out in Europe by Kalman Tihanyi, J. D. McGee, and others. Nonetheless, RCA contended that the Iconoscope was essentially the same device that Zworykin tried to patent in 1923 and that it was virtually the same invention as the Image Dissector. RCA conveniently ignored the fact that Zworykin was still working with spinning disks and mirrors all through the late twenties, right up until the time he visited 202 Green Street. Nonetheless, it laid the foundation for a legal claim that the inventor of television was not Farnsworth but Zworykin. As a result, Sarnoff said that he didn't have to pay any royalties to Farnsworth for the right to manufacture television sets. "RCA doesn't pay royalties," he is alleged to have said. "We collect them."

A rancorous legal dispute inevitably ensued over who had actually invented television. RCA's lawyers contended that Zworykin's 1923 patent took priority over any of Farnsworth's patents, including the one for his Image Dissector. However, RCA's position was fairly weak because it could furnish no evidence that in 1923 Zworykin had produced an operable television transmitter. Their case suffered a further setback when Farnsworth's old chemistry teacher, Justin Tolman, stepped forward to testify that Farnsworth had conceived the idea when he was a high school student. He even provided the court with the original sketch of an electronic tube that Farnsworth had drawn for him at that time. The drawing was almost an exact rendering of the Image Dissector Farnsworth had gone on to invent.

Ruling on the case in 1934, the U.S. Patent Office awarded priority of invention to Farnsworth. RCA appealed and lost. But with its deep pockets, the company was able to draw the courtroom battles out for several years until Sarnoff finally agreed to pay Farnsworth royalties. The four years of legal challenges succeeded in slowing the development of television, delaying its introduction to the public. For Farnsworth, the battle took its toll both in terms of his finances and his health.

In spite of the controversy surrounding the patents, television finally did make its public debut. By 1935 regular broadcasting service had begun in Germany, with medium-definition images of 180 lines. To establish its association with television in the minds of Americans, Sarnoff masterminded a public relations coup. In 1939, RCA took advantage of the World's Fair in New York City's Flushing Meadows to sponsor the Television Pavilion, managing to secure the rights to host and broadcast the opening ceremony on radio *and* television. Capitalizing on the ensuing publicity, he also made certain that New York department stores were stocked with newly minted RCA models.

But the commercial development of television under the RCA label quickly came to a standstill with the outbreak of World War II, which brought about the suspension of all commercial television production and sales. What might at first have seemed a setback for RCA actually worked to its advantage. By the end of the war, Farnsworth's key patents were close to expiring. As soon as they did, RCA rushed in to fill the void and took over the production and sales of television sets. In 1947, when the patents expired, RCA was producing six thousand televisions a year; by the mid-1950s it was shipping millions, capturing nearly 80 percent of the U.S. market. At the same time, the corporate giant instigated an aggressive public relations campaign, promoting both Zworykin and Sarnoff as the fathers of television. With no resources to fall back on, Farnsworth sold the assets of his company to International Telephone and Telegraph, which shortly afterward decided to get out of the television business altogether. Devastated, Farnsworth suffered a nervous breakdown that left him bedridden for months.

While Farnsworth was later to work as a consultant in electronics and a researcher in atomic energy, he never seemed to recover from the succession of blows dealt to him by RCA. The final years of Farnsworth's life were marred by tragedy. Disillusioned, he withdrew with his family to a house in Maine, where he was tormented by depression and alcoholism. In 1947, his house burned to the ground. Ten years later, he was so unknown to the public that he was tapped to be a mystery guest on the television program *What's My Line?* Farnsworth was introduced as Dr. X. It was up to the panel to ask him questions and, by his answers, determine what he had done in his life to merit his appearance on the show. One of the panelists asked Dr. X if he had invented some kind of a machine that might be painful when used. Farnsworth answered, "Yes. Sometimes it's most painful."

Farnsworth died in 1971, the same year as Sarnoff.

CHAPTER 7

A Faint Shadow of Its Former Self

Alexander Fleming and the Discovery of Penicillin

For millennia mankind was at the mercy of microbes. In China, texts dating back to the thirteenth century B.C. document the fear of an outbreak of infectious disease. "Will this year have pestilence and will it be deaths?" the ruler of an ancient Chinese kingdom asked the diviners. The Black Plague of the Middle Ages wiped out a third of Europe's population. When the plague was raging, a person could be in robust health one day and keel over and die the next. "A world view allowing ample scope to arbitrary, inexplicable catastrophe alone was compatible with the grim reality of plague," writes William McNeil in his classic study *Plagues and Peoples.*

It wasn't until late in the nineteenth century that the state of medical knowledge had advanced enough for scientists to begin to think that perhaps these catastrophes could be explained, after all. And further, if their causes could be understood, it might be possible to put a stop to them. In the 1870s, Robert Koch (1843–1910), a German scientist and future Nobel laureate, succeeded in isolating several disease-causing bacteria, including those of tuberculosis. Moreover, he was able to identify many of the animal vectors for these diseases. (Vectors are organisms—insects and plants as well as animals—that carry diseases; they are not necessarily the source of the disease.) Koch's most momentous contribution to modern medicine was the isolation of the anthrax bacillus, a disease-causing organism. This was the first time that the causative agent of an infectious disease had been demonstrated beyond a reasonable doubt. It was now clear that infectious diseases were not caused by mysterious substances, but rather by specific microorganisms. Anthrax, for instance, was caused by a particular type of bacterium. Koch also established the working principles for studying these microorganisms that are followed by researchers to this day.

Koch's germ theory of disease gave doctors a far better understanding of how infection is caused. What they were still missing, however, was a better way of treating these infections. Their arsenal was basically limited to vaccinations and antitoxins. A carbolic spray, developed by British surgeon Joseph Lister in the latter part of the nineteenth century, worked with some success during operations, but was not suitable for everyday use because it burned tissue. The question that doctors and scientists now sought to answer was, How could people be protected against infection on a day-to-day basis?

The question was given greater urgency with the outbreak of the First World War. On the Western Front more wounded soldiers were dying from infections than were being killed outright in the trenches. Because the infections were relatively simple, they should have been relatively easy to cure—if only some chemical or compound could be found that was capable of fighting them.

The problem was of particular interest to a Scottish bacteriologist assigned to a battlefield hospital laboratory in France. He had come recommended to his superiors on the basis of his experience as a private practitioner in London. There he had established a prosperous sideline treating prominent artists with syphilis using a new compound called salvarsan (which means "that which saves by arsenic"). It occurred to the bacteriologist that something similar could be found that would be equally useful against infections that were proving to be so devastating in the trenches. The bacteriologist's name was Alexander Fleming.

A modest little man with shaggy hair and spectacles who favored bow ties, Fleming had always been a bit of an eccentric, and even when he was a celebrity he had trouble mastering the conventions of society. But perhaps it was his failure to follow convention that accounted for his instinctive ability to see what others tended to overlook.

Fleming was born in a remote, rural part of Scotland in 1881, the seventh of eight siblings and half siblings. Raised on an eight-hundred-acre farm a mile from the nearest house, he spent much of his childhood roaming the valleys and moors of the countryside. For the future scientist, the experience was a formative one. "We unconsciously learned a great deal from nature," he later observed, referring to himself and his brothers.

At the age of 14, Alec, as he was known, followed his brother Tom to London, where he enrolled in the Polytechnic School. Tom encouraged him to enter business, but his one try at it—working in a shipping firm—failed to convince him that he was cut out for a life sorting through bills of lading. When the Boer War broke out between the United Kingdom and its colonies in southern Africa, Alec and two of his brothers joined a Scottish regiment. Although no one in the regiment ever saw any action, they had more than ample opportunity to go swimming and practice their shooting skills. Fleming's marksmanship would hardly be worth noting except that in a roundabout way it influenced the trajectory of his future career.

When he returned to London, Fleming announced his intention to study medicine. It is likely that he was inspired by the example of his older brother, who had already established a thriving medical practice. After earning top scores in the qualifying examinations, Fleming had his choice of three medical schools. Knowing little about them, he chose St. Mary's simply because he had once played water polo against its team. Almost arbitrarily, he decided to specialize in surgery. This was a source of consternation to a friend of his who was also head of the rifle club. If Fleming was to go ahead with his plans to become a surgeon, eventually he would have to leave St. Mary's to take up a position in a hospital, thereby depriving the club of its crack shot. As it turned out, there was an alternative. The head of the rifle club worked in the Inoculation Service; if Fleming wanted, he could join the service, too, and stay on at St. Mary's. To do so would also allow Fleming the opportunity to study under Almroth Wright (1861–1947), a brilliant researcher who had discovered a typhoid vaccine in 1896. Fleming wasn't hard to convince. Except for his service in the Royal Army Medical Corps during the war, Fleming would stay at St. Mary's for the rest of his career.

After the Armistice in 1918, Fleming returned to research and teaching at St. Mary's, where he concentrated most of his research on developing an effective antiseptic. In the process he discovered lysozome, an enzyme found in many body fluids, such as tears. Although it had natural antibacterial properties, it had little effect against the strongest infectious agents. He kept looking, pursuing a pet theory of his—that his own nasal mucus had antibacterial effects. By the late summer of 1928, though, his attention had largely turned to the study of staphylococci. Staphylococci are an especially pernicious type of parasitic bacteria commonly found in air and water that are capable of causing pneumonia and septicemia as well as boils, and kidney and wound infections. Always a little disorganized, Fleming had so much going on in his lab that it was often in a jumble. As it turned out, the disorder would prove fortuitous.

It was French chemist Louis Pasteur (1822–1895) who famously noted that "chance favors only the prepared mind." In the history of science there are few people who better exemplify this dictum than Alexander Fleming. Echoing Pasteur, Fleming once said, "Do not wait for fortune to smile on you; prepare yourself with knowledge." In some way, everything that Fleming had done up to this point can be considered just such a preparation.

In the waning days of the summer of 1928, Fleming left London on holiday. For whatever reason, he failed to store his cultures of staphylococci in incubators where they would be kept warm. Instead he left the cultures in petri dishes out in the open. Here some knowledge of the lab's layout is helpful. Because it was almost impossible for him to open his window, Fleming usually left the door open to get some air circulating. The door opened onto a flight of stairs. One flight down was another lab that was being used by a young Irish mycologist named C. J. La Touche, whose door opened onto the same flight of stairs. La Touche was working with a strain of mold that would turn out to have some very interesting properties. But his laboratory lacked a fume hood, a kind of isolation chamber. Had La Touche used such a hood, he would have been able to confine the spores to a small isolated area. Instead the spores from the mold spread throughout the mycologist's lab and then drifted out the open door and up the stairs until they found their way into Fleming's laboratory.

The weather abetted fortune, too. During Fleming's absence, London was hit by an unusually cold spell, followed immediately by a return of warm temperatures, a cycle that caused the spores from La Touche's lab to flourish in their new home upstairs.

When Fleming returned from his holiday in September, he began to discard some petri dishes that he'd left out in the open. Ordinarily, contamination is the bane of bacteriological work. Bacteriologists look on contaminants in much the same way that farmers regard weeds. When a culture is contaminated, the first instinct of a scientist is to throw it away and start fresh. And that was exactly what Fleming did in a kind of routine housecleaning. But once again serendipity intervened. At the time, the Inoculation Department of St. Mary's used disposal containers consisting of shallow enamel trays with a little antiseptic. (If Fleming's had been a properly equipped bacteriological lab, he would have had deep buckets filled to the brim with antiseptic.) It was only then, having discarded the contaminated cultures, that he observed a clear halo surrounding the yellow-green growth of mold that had accidentally contaminated the plate. Of course, at this point he had no way of knowing that a spore of a rare variant of a mold called *Penicillium notatum* had drifted in from the mycology lab one floor below.

A number of events had to have occurred in order to produce the phenomenon that Fleming observed. First of all, the staphylococcal cultures were left exposed rather than stored in a warm incubator, where they would never have been contaminated by La Touche's spores. Moreover, the unexpected cold spell gave the mold a chance to take root and grow. Later, when temperatures rose, the staphylococci were able to flourish, spreading like a lawn until they covered the entire petri dish—except for the area directly exposed to the moldy contaminant. "It was astonishing that for some considerable distance around the mould growth the staphylococcal colonies were undergoing lysis (the dissolution or destruction of cells)," Fleming wrote. "What had formerly been a well-grown colony was now a faint shadow of its former self." This lysis, or destructive process, he realized, was what was responsible for discoloring his microbes. He correctly deduced that the mold must have released a substance that simultaneously destroyed existing bacteria and inhibited their further growth. Literally pulled from the garbage, it was a discovery that would change the course of history.

It took a perspicacious scientist like Fleming to observe the unusual halo effect in the first place and then to recognize its significance. This was Fleming's eureka moment, a result of a combination of personal insight and deductive reasoning. His long years of preparation had paid off.

Fleming hadn't discovered *Penicillium.* The existence of this genus of molds was hardly a mystery and its effects were well known. The mold accounts for the green streaks in Roquefort cheese, for example, and if it spreads into an orchard it can quickly spoil fruit. In fact, Fleming wasn't even the first to describe the antibacterial properties of penicilliums. In 1871 the English surgeon Joseph Lister noticed by chance that the mold was able to sharply curtail the growth of germs. He went so far as to carry out some successful experiments on patients, but never seemed to entirely grasp the importance of his findings. Two other researchers—John Tyndall in 1875 and D. A. Gratia in 1925—had also noted the mold's antibacterial action. But like Lister, they hadn't seemed to appreciate the significance of their observations, nor did they conduct the necessary experiments to find out exactly why the mold killed bacteria. One of the reasons that science does not always move in some kind of inexorable march of progress is that discoveries often get overlooked or discredited because they fly in the face of conventional wisdom. To some extent, this is what happened in the case of penicillin. When the world finally learned of the therapeutic properties of penicillin, a doctor was prompted to write a letter to the *New England Journal of Medicine* recounting his experience with the mold. The author had attended medical school in the late 1800s, a time when vast numbers of patients with bacterial infections could expect to die. In bacteriology class, he had carelessly allowed his culture dish to become contaminated with mold. He noticed that the mold killed the bacteria. Far from praising the student for his powers of observation, or even expressing curiosity, his professor had berated him for his sloppiness in allowing the contamination to occur. Because the professor was exclusively focused on obtaining the "correct" result, neither he nor his student asked the proper question, namely, "Can the property of molds to kill bacteria in vitro (in a culture) be used to cure bacterial infections in vivo (in a living organism)?"

Fleming had learned another crucial lesson in science—part of his "preparation of knowledge"—and that lesson was that it is often just as important to know which questions to ask as it is to look for answers. He was characteristically modest in assessing his role in the discovery: "My only merit is that I did not neglect the observation and that I pursued the subject as a bacteriologist."

What was also apparent to Fleming was that it wasn't sufficient to make a remarkable discovery. If it were to be of any practical use, he would also have to investigate the substance that he had so fortuitously stumbled upon.

Never one given to overheated prose, he noted simply, "I was sufficiently interested to pursue the subject."

The most-used tool of a microbe hunter is a pretty simple instrument, consisting of nothing more than a loop attached to the end of a platinum wire. Using this device, Fleming proceeded to scoop up a speck of the mold from the colony of bacteria and placed it in peptone broth, the food molds feed on. Then as the mold grew, he examined it under his microscope. At first it had a fuzzy white appearance, but as the days went on, it turned into a tufted green mass. His microscopic examination provided him with the final proof that the mold belonged to a large genus of molds called *Penicillium* (from the Latin meaning "little brush").

Fleming assumed that the mold was brewing a juice that killed his staphylococcal colonies. This meant that he would have to look for it in the broth in which it was flourishing. So he filtered off some of the mold and dropped some of the strain on the glass plate in which his healthy staphylococcal colonies were growing. After several hours the bacteria died, disappearing right before his eyes as he watched through the lens of his microscope. His next step was to find out how little of the substance was necessary to kill off the bacteria. So he began to dilute the mold-containing broth. When he tried it out at one hundredth of its original strength, he was delighted to find that it still worked as well as before. He continued to dilute it further and further, but even when he had diluted the broth to eight hundredths of its original strength, the mold still retained its lethal power against the bacteria. The miraculous substance had an added advantage in that it was several times as potent as pure carbolic acid, which, while killing bacteria, also burns the tissues. Fleming and his colleagues repeated the procedure with pneumococci, the bacteria that cause pneumonia, and produced the same astonishing results.

He wondered what would happen if he exposed some of his own saliva to the penicilliums. Since saliva is full of all kinds of bacteria, he theorized that it should have a pronounced effect. He placed the specimen of saliva on a fresh plate of jellylike agar and put it in an incubator. As expected, the colonies of assorted bacteria quickly grew in profusion. Then he added penicillin. Some colonies were wiped out; others continued to thrive. Those that were destroyed were evidently sensitive to penicillin. Other colonies proved impervious to the mold. The drug, it seemed, was effective against some types of bacteria and not others. When he published his results in 1929, Fleming suggested that penicillin could be used as a helpful laboratory tool to separate the "goats from the sheep," as it were, in a mixed culture. "It has been demonstrated that a species of *Penicillium* produces in culture a very powerful antibacterial substance," Fleming noted. "It is a more powerful inhibitory agent than carbolic acid and it can be applied to an infected surface undiluted as it is non-irritating and non-toxic."

In other experiments, he found that while penicillin was ineffective against bacteria that caused typhoid fever, dysentery, and certain intestinal infections, it worked very well against the bacteria that caused pneumonia, syphilis, gonorrhea, diphtheria, and scarlet fever.

Even though the penicillium broth had proven itself in the laboratory, the question arose as to whether it would work as well in human beings. But first Fleming had to demonstrate that it was safe. To test for toxic effects, he tried the broth out on mice and rabbits, injecting a thimbleful of the broth in their ears through a syringe. But the mice and rabbits showed no ill effects.

By 1932, Fleming had abandoned his work on penicillin. Although his name would forever be associated with its discovery, he would have no further role in investigating or developing the antibiotic. However, he made certain to safeguard the unusual strain of *Penicillium notatum* for posterity and continued to make samples of the mold available to other researchers. His decision appears to have been motivated at least in part by a lack of funding and chemical expertise to purify the penicillin. For while he had experimented with the broth, he had never tried to separate out the specific microbe-destroying elements that made it work so well against bacteria. Some historians speculate that he might even have had some doubts about whether the drug could cure serious infections in spite of compelling evidence that it did.

But Fleming's withdrawal could hardly dampen the enthusiasm he had stirred up by his startling discovery. Scientists now had to find some way of turning the penicillin mold broth into a pure drug that would be more suitable for the treatment of infections in human beings.

In 1935 a German scientist named Gerhard Domagk was able to show that the injection of a simple compound, prontosil, could cure systemic streptococcal infections. This was the first time that invading bacteria had been killed with a drug. This breakthrough inspired a fevered search for similar compounds, the most promising of which was *Penicillium notatum.*

There are actually several kinds of penicillin synthesized by various species of the mold. Regardless of how they are produced, though, all penicillins work in the same way, namely, by inhibiting the bacterial enzymes responsible for cell-wall synthesis and by activating other enzymes to break down the organisms' protective walls. As a result, they are not effective against microorganisms that lack cell walls.

The next chapter in the penicillin story belongs chiefly to Howard Florey and Ernst Chain. Born in 1898 in Australia, Howard Florey had practiced as a doctor before going to Oxford University, where he was appointed professor of pathology. Florey had already made several important scientific discoveries, but his main focus was directed toward natural substances that could kill bacteria. He had been intrigued by the possibilities of penicillin ever since he read an article by Fleming published in 1929, a year after the drug's discovery. But it was ten years later that Florey's team at Oxford actually got a sample of the mold to work with. Unlike Fleming, Florey had a large research department and a rich supply of talent. His principal collaborator was Ernst Chain, a brilliant German-Jewish chemist who had fled persecution in Nazi Germany. With backing from the Rockefeller Foundation, the two scientists and their colleagues set out to identify and isolate substances from molds that could kill bacteria in order to purify the penicillium broth into a drug. There was also a related question as to whether the drug, assuming they could make it, could be produced in sufficient quantities, and cost effectively, for widespread use.

Florey, Chain, and their colleagues rapidly purified penicillin in sufficient quantities to perform the experiment that Fleming could not, and they successfully treated mice that had been given lethal doses of bacteria. Within a year, their results were published in a seminal paper in the prestigious British medical journal *Lancet.* By 1940, Florey and Chain were confident enough of the drug's efficacy and safety that they were ready to try it out on their first human patient, a policeman named Albert Alexander who was suffering from a dangerous bacterial infection. The patient began to recover after receiving the drug. All the same, the scientists were constrained from duplicating their tests on humans on any large scale because they couldn't produce enough of the drug. This situation rapidly began to change in the early 1940s, when Florey succeeded in galvanizing several major pharmaceutical companies into helping him, ensuring that the Oxford researchers would have the cash and resources necessary to continue with their experiments on humans. By 1942, because of treatment with penicillin, several people were alive who would otherwise have succumbed to their infections.

But it was World War II that was most responsible for making penicillin a household name. Here, at last, was a miracle drug that had the capacity to drastically cut down the number of battlefield deaths caused by infected wounds. When the United

States entered the war in December 1941, Florey managed to persuade American drug manufacturers that it was in their best interests to mass-produce penicillin. Like the Manhattan Project, which developed the atomic bomb, the U.S. government threw its weight behind the effort, providing grants to drug companies to defray some of the costs in acquiring the expensive equipment needed to make penicillin. Soon, so many companies were scrambling to get in on the action that penicillin had become big business. To coordinate the production effort, the government and drug companies established a network of "minifactories" dedicated exclusively to turning out batches of penicillin as quickly as possible. In 1943 the British pharmaceutical firms geared up their own mass production of the antibiotic.

A tireless crusader, Florey sought to bring penicillin to every corner of the globe. To that end he traveled to Russia in 1943 to help the Soviets set up their own production capacity. By D Day, there was enough penicillin on hand to supply every soldier who needed it in every Allied army. By the end of World War II, it had saved millions of lives. And as penicillin became more widely available, its price began to drop, putting it in reach of the populations of impoverished and war-devastated countries. And it was proving to be useful in treating a wider variety of ailments than even Florey or Chain had imagined, including throat infections, pneumonia, spinal meningitis, gas gangrene, diphtheria, syphilis, and gonorrhea, as well as life-threatening infections to mothers in childbirth. Long and complex surgical procedures could now be performed with much greater assurance that the patient wouldn't die of a postoperative infection.

Penicillin was by no means the only drug that had been developed to combat infections, though it was far and away the best known. The first antibiotic to be used successfully in the treatment of human disease was in fact tyrothricin, which was isolated from certain soil bacteria by American bacteriologist René Dubos in 1939. While it can be applied externally for some infections—on the skin, for instance—tyrothricin is too toxic for general use. Scientists also discovered another class of antibiotics produced by a group of soil bacteria called actinomycetes. These proved safer to use and more successful. One of these was streptomycin. Discovered in 1944 by the American biochemist Selman Waksman and his associates, the drug remained for years the major treatment for tuberculosis.

But no drug could rival penicillin in terms of a good story. And what a story it was, too, with all its attendant trappings of serendipity: the drifting spores, the open doors, the puzzling halo in the culture dish. The account of Fleming's discovery had begun to assume the aspect of a myth. And for the story to have resonance, it could only have a single hero. That hero, naturally, was Alexander Fleming. Although he shared with both Florey and Chain the 1945 Nobel Prize in medicine for his work with penicillin, it was Fleming alone whom most people would associate with the discovery of the drug. Even though the prize was given to all three scientists, disagreement quickly broke out about who should get the major credit for the development of the antibiotic.

Nor was Fleming shy about seeking public adulation. On the contrary, he embraced it, reveling in the publicity. He was happy to court the press, while Florey would have nothing to do with reporters. Florey's reticence accounts in part for why his contribution never received the attention that Fleming's did. Moreover, it is obviously easier for people to identify with a lone scientist experiencing an epiphany in his lab than with a team of persevering researchers conducting hundreds of painstaking experiments over a period of years. A host of awards and honors were heaped on Fleming, including a knighthood (which he shared with Florey) in 1944.

But Fleming was farsighted enough to realize that the widespread use of penicillin—and many of the wonder drugs that were to follow—posed a hidden danger. Writing in 1946, he warned that, "the administration of too small doses . . . leads to the production of resistant strains of bacteria." Even so, with the introduction of more potent antibiotics throughout the 1950s—Fleming died in 1955—few people paid much attention to such dire predictions. Some commentators pictured a world that was entirely free of infectious disease.

Unfortunately, Fleming's words proved all too prophetic. Some strains of previously susceptible bacteria have developed a specific resistance to penicillin; these bacteria either produce penicillinases, enzymes that disrupt the internal structure of penicillin and thus destroy the antimicrobial action of the drug, or they lack cell-wall receptors for penicillin, greatly reducing the drug's ability to enter bacterial cells.

Today, the phenomenon of bacterial resistance has reached alarming proportions, to the point where recently even the most powerful antibiotics have proven ineffectual in many infections. There are several reasons to account for this. For one thing, people have overused and misused antibiotics, taking them for minor afflictions or for illnesses that aren't bacterial in origin at all, like the flu, which is caused by a virus. Or else people may try to economize and stretch out their doses far beyond the recommended protocol. Others stop taking their medication when the symptoms disappear but before the bacteria are eliminated. This means that they aren't getting an effective dose at any given time, allowing the stronger, or more resistant, members of a bacterial colony the opportunity to flourish even as the weaker bacteria are killed off. Or else people take an antibiotic they've kept in the medicine cabinet for months or years, long past its expiration date. Some physicians compound the problem by overprescribing antibiotics to satisfy the demands of their patients.

To make matters worse, livestock are routinely treated with massive doses of antibiotics—a situation that has been going on for decades—both to prevent the spread of disease in quarters where the animals are closely packed together and to promote growth. (No one is precisely certain why antibiotics have this effect.) This means that consumers may be absorbing antibiotics when they sit down to eat a hamburger. As a result, more and more Americans are unwittingly developing resistance to ever more antibiotics, putting themselves at risk for a time when they are truly in need of the drugs to fight dangerous infections. It is almost a classic illustration of the cliché: too much of a good thing.

We now face the prospect of seeing Fleming's legacy undone by the excessive use to which antibiotics have been put ever since wandering spores of mold first blew into his lab in the summer of 1928. It is one of those ironic twists of fate, which often mark the history of science, that advances once hailed as miraculous lead to unforeseen consequences, which in some cases, carry with them the seeds of their own destruction.

CHAPTER 8

A Flash of Light in Franklin Park

Charles Townes and the Invention of the Laser

Lasers are now so widely used that they have become commonplace. In scientific research, they have provided new insights into our understanding of the nature of light, while in industry they are routinely used in communications systems, precision welding and drilling into heat-resistant materials, and measuring with a high degree of accuracy. Inevitably, lasers have been seized upon by the military to design cutting-edge navigational and weapons-targeting systems. And in medicine, lasers are now showing promise in treating cancer by destroying diseased cells. Lasers also threaten to put many optometrists out of business because of their ability to weld detached retinas and correct myopia without the need for invasive surgery. Lasers are widespread in the entertainment industry: CDs and DVDs wouldn't exist without them. And lasers might one day unlock one of the great riddles of the universe by detecting gravitational waves, which Einstein predicted but which have thus far not been found.

Taken for granted now, lasers were once regarded as a solution in search of a problem. Until it was invented and proven practical, there was no apparent need or desire for the laser. The idea of using concentrated beams of light as a power source simply wasn't something that struck most people (including scientists) as an especially promising idea. Moreover, the one scientist who is arguably most responsible for conceiving of the laser had no intention of inventing such a thing. On the contrary, the laser was something of an afterthought, a potentially interesting elaboration on an instrument designed to amplify the strength of microwaves. To understand the genesis of lasers, you first need to understand something about microwave technology. For that reason, the kitchen is a good place to start.

It is unlikely that many famished Americans who rely on microwave ovens to heat their food in a matter of minutes give much thought to just how these devices produce heat. And how many harried homemakers would be able to define exactly what a microwave is? The very ubiquity of microwave technology—it has applications in radio and television, radar, meteorology, satellite communications, and distance measuring—has robbed it of much of its mystery. But in fact, the notion that microwaves could one day be useful for detecting enemy aircraft or heating a cheeseburger never even entered scientists' minds when they first began to investigate the phenomenon. Their interest in the subject was almost entirely theoretical. Microwaves, they believed, could be harnessed as an important tool to study the properties of matter.

Microwaves are short, high-frequency radio waves lying roughly between very-high-frequency (infrared) waves and conventional radio waves. (Specifically, they range in length from about 1 mm to 30 cm, or about 0.04 to 12 inches.) Microwaves, it turns out, have the capacity to stimulate atoms, allowing scientists to investigate matter in different states. This is the principle behind the humble microwave oven. The microwaves enter the oven through openings in the top of the cooking cavity, where a stirrer scatters them evenly throughout the oven. Once the microwaves penetrate the food (assuming that it is not protected by metal) they agitate the water molecules in the food, causing them to vibrate, which in turn produces heat. In 1916 Albert Einstein (1879–1955) had shown theoretically that atoms stimulated by radiation could emit, as well as absorb, radiation. At the time, microwave technology was based mainly on systems and devices developed from the work in the early 1900s of Ernst Alexanderson (1878–1975) of General Electric, Lee De Forest (1873–1961) of New York, and John Ambrose Fleming (1849–1945) of London. By the late 1930s, most microwave technology relied on vacuum tubes capable of emitting microwaves as short as a few millimeters.* If, however, the wavelengths could be made even shorter, scientists reasoned, then they could produce stronger radiation. This would offer them a better understanding of how the action of molecules and atoms could be used to control radiation. But this could not be accomplished so long as the technology was limited to vacuum tubes. Some other devices would have to be invented to generate waves short enough to achieve the effect researchers wanted. Among these researchers was a young South Carolina physicist named Charles H. Townes.

When Townes set out to develop a more efficient device for generating shorter microwave radiation, he only intended it as an aid in the study of molecular structures. He certainly wasn't thinking

* Vacuum tubes, consisting of electrodes—metal plates—and wires in an evacuated glass bulb, are capable of regulating electric currents or electronic signals. Until the invention of the transistor in 1948, vacuum tubes were used not only to generate microwaves but in televisions, radios, and early computers as well.

about inventing a revolutionary instrument. But that is exactly what he ended up doing.

Charles Townes began his life on July 28, 1915, on a farm in a mountainous region of South Carolina not far from Greenville. Growing up on a farm was an invaluable experience for a budding scientist. "I liked to figure things out and see how they worked and try to make them work," he told an interviewer many years later. "So I made all kinds of things—there were wagons and slingshots, and so on. And tried to make them better and bigger." When he was 10, he told his sister that whatever she bought him for Christmas she should be sure to buy it at a hardware store. His father, a lawyer, encouraged his enthusiasms, using the natural setting of the farm as a teaching tool. He brought the young Charles old clocks so that he could take them apart and see how they worked, then put them back together. He tinkered with automobiles and small radios, whatever would give him the opportunity to innovate. "You find that if you don't have exactly the right tools, you don't have exactly the right batteries or something, you can think, 'Well now, how can I do this otherwise and fix it?' " All the same, because of his love of animals and nature, Charles assumed that he would become a biologist. But no sooner was he exposed to physics at Furman University than he changed his mind. (Physics wasn't taught at his high school.) "I said, oh, physics! That really figures things out thoroughly, it tells you exactly what's likely to be right. And that was very pleasing." Excited, he told a friend about his ambition. His friend had no idea what physics was. Charles likened it to chemistry and electrical engineering, subjects which were more familiar.

By the time that Townes moved west to do his graduate work at California Institute of Technology in the mid-1930s, physics was no longer banished from high school curricula. For a budding physicist, a headier time could hardly be imagined. Quantum mechanics was coming into its own, and nuclear physics, as Townes put it, "had just been sort of discovered." Caltech was an especially exciting place to be; Townes took courses with both Robert Oppenheimer, who later became famous for his work on the development of the atomic bomb, and the Nobel laureate Robert Millikan, best known for his work in atomic physics. But in 1939, the year Townes was finishing up at Caltech, the Depression had yet to relinquish its grip on the country and there were not a lot of jobs available for a physicist with a Ph.D.

Townes's original intention was to teach at a university and do research. He had no desire at all to go into private industry. Even after he was interviewed and offered a job at Bell Laboratories in New Jersey, he was still hoping to find an alternative. One of his professors urged him to reconsider, reminding him that Bell Labs had established a reputation for good science. Reluctantly Townes agreed to accept the job, a decision in hindsight he was lucky to have made.

Bell Labs allowed Townes to work on a variety of problems, including microwave generation, vacuum tubes, magnetics, and solid-state physics, which gave him a chance to study how electrons are emitted by surfaces. He wasn't allowed the luxury of researching basic physics for long, though. "All of a sudden, the big boss at Bell Labs (Mervin Kelley) called me in his office and said, 'On Monday you are to start working on radar systems for navigation and bombing.' " Townes wasn't pleased. He would have preferred to focus on the nascent field of radio astronomy. Besides, he realized that if he were to do what Kelley wanted, he would have to become an engineer as well as a physicist. On the other hand, it was as clear to him as to everyone else in the country in 1939 that war was approaching and scientists would have to do their part in trying to help the country gear up for it. Once again he discovered an unexpected benefit by being forced from the path that he would have ordinarily chosen. "I learned a lot of new things which many physicists didn't know," he remarked. One of the "new things" was electronics. Another was microwaves.

In 1939, two British scientists—Henry Boot and John Randall—had developed a device known as a resonant-cavity magnetron, which was capable of generating high-frequency radio pulses with a great deal of power, setting the stage for the development of microwave radar. The basic principle of radar is relatively straightforward. Radar systems broadcast radio signals at specific wavelengths. When these signals strike solid objects, such as a battleship or an airplane, the signals that reflect back to the radar system can be correlated to identify the object and its position. The smaller the waves, the better the information received. The main problem comes in trying to accurately pinpoint and achieve a clearer resolution of the target. The radar navigation bombing systems Townes was working on used wavelengths first of 10 cm and then 3 cm, but the military wanted to reduce the length of the wavelength further, to 1¼ cm, which would allow radar to more precisely locate the target and which would have the added advantage of requiring smaller antennas on the planes. The chief stumbling block, as Townes knew, was that at the desired wavelength, gas molecules can absorb the wave forms. This meant that water vapor in the atmosphere—fog, rain, and clouds—could prevent the radar from performing efficiently. Nonetheless, Townes and his team built the radar and tried it out. The results were just as Townes had feared. "The system could 'see' for just a few miles at best," he said. In spite of the disappointment, Townes didn't abandon his interest in microwave spectroscopy—far from it. He said, "You follow a path, it's a path nobody else has followed. You are an explorer . . . you never know just what you're going to find." He argued successfully that his employer should allow him the freedom to work on the interaction of microwaves and molecules because of the importance of microwaves to communications.

After the war, Townes was invited to accept a research position at Columbia University. Townes was happy to accept, both because Columbia offered him the opportunity to pursue the research that was most meaningful to him and because he preferred an academic environment to a corporate one. Moreover, he brought to the university some unique skills from his experience at Bell Labs working in an industrial laboratory with engineers. Physicists, after all, didn't have much to do with devices like amplifiers and oscillators whereas engineers did. On the other hand, engineers didn't generally know the first thing about quantum mechanics and atoms and molecules, which were the dominion of the physicist. Physicists, however, were locked in by a kind of mind-set, as Townes put it. Problems involving amplifiers and oscillators simply didn't interest them very much. What Townes could bring to the table was an ability to draw on both disciplines—what he called a "marriage of physics and engineering, the marriage of quantum mechanics and electrical engineering."

At Columbia, Townes concentrated his efforts on microwave spectroscopy, trying to find a solution to the problem that had proven so vexing at Bell Labs: how to achieve shorter waves. He was looking for some device that would generate microwaves in great intensity—the shorter the wave, the stronger it is. Everything he tried, failed to work, but undaunted, he continued to mull over the problem for several years. Then in 1951, four years after arriving at Columbia, Townes hit upon the idea of producing this energy not by electronic circuits, which had been the focus of his earlier efforts, but rather by manipulating the molecules themselves. The question was how to accomplish this.

Townes immediately thought of the ammonia molecule, which is a very strong absorber and interacts strongly with wavelengths. He characterized ammonia as "an old favorite of mine" that he'd studied a great deal in the past. He hypothesized that he would be able to get the ammonia molecules "excited" by pumping energy into them through heat or electricity, after which he would expose them to a weak beam of microwaves. Molecules excited in this way would then be impelled to emit their own energy in microwaves, which would bombard other molecules in turn in a kind of domino effect, causing them to give up their energy. By using the very feeble incoming microwaves generated by ammonia molecules, he hoped to initiate a cascade of chemical processes that would produce a highly amplified beam of radiation.

Among his colleagues at Columbia his idea barely stirred any interest—which to Townes's mind was all to the good. It allowed him the freedom to work with his graduate students without any hindrance. "We could work along peacefully on it in a graduate student kind of style for three years, and we finally made it work and nobody else was trying to do it that I knew of." People would wander into his laboratory to observe his work in progress. They all agreed that it was a very nice idea and then went away. "They didn't think it was worthwhile, worth the effort," he said. "But I thought I saw some possibilities there."

In December 1953, Townes's team succeeded in constructing a device that would produce strong microwaves in any direction. They called the process microwave amplification by stimulated emission of radiation, which became known more popularly as the maser. The maser quickly found many applications for its ability to send strong microwaves in any direction. The first masers used a static electrical charge to discard low-energy molecules of ammonia gas which excited the first batch of molecules inside a cavity; as those molecules returned to a lower energy level, they released microwave radiation. This radiation, reflecting inside the cavity, stimulated additional molecules to radiate energy. This process amplified, or intensified, the microwave radiation emitted by the device.

From the start, the maser was a big hit, resulting in improvements in radar and in amplifying radio signals. It would also find uses as an ultrasensitive instrument for detecting and measuring radiation in space. In addition, the new device provided the basis for an atomic clock that was far more accurate than any mechanical timepiece ever invented.

In spite of the success of the maser, Townes still wasn't satisfied. If the maser could do so well at producing microwave amplification, he thought, why couldn't a similar instrument do the same with beams of light? But why light at all? The study of light was an old field. "Even when I was a student in the 1930s," Townes said, "light was considered to be understood and everybody knew about light and there wasn't anything much new to come out of it." That is, while light was exceedingly useful and used so often that everyone took it for granted, few people believed that there were any possible applications for it that hadn't already been achieved. Townes, however, was of the opinion that there was more to be learned about light. Even so, he still wasn't trying to create anything practical. He said, "The reason I was trying to get short waves was to do science, find out more about molecules. I wasn't looking for an application. I wasn't thinking of a laser beam that would be a bright light or something. I wanted to find out more about molecules and I wanted to get shorter waves to study the molecules. Just basic work, not applied at all."

In grappling with the question, he sought the help of another physicist who shared Townes's fascination with microwave spectroscopy: Arthur Schawlow (b. 1921). Schawlow and Townes had first met in 1949 at Columbia. The two had collaborated in the past and coauthored a book about microwave spectroscopy that appeared in 1955. By the mid-1950s, the two also had something else in common aside from their scientific interests: Townes had married Schawlow's sister.

In 1956, while still at Columbia, Townes was offered a consulting position with Bell Labs, which he accepted. It so happened that Schawlow was working for Bell Labs at the time, and one day Townes dropped in on him at his lab to discuss the direction his thoughts were taking him. He found a receptive audience. "I was beginning to think seriously about the possibility of extending the maser principle from the microwave region to shorter wavelengths, such as the infrared region of the spectrum," Schawlow later recalled. "It turned out that he (Townes) was also thinking about this problem, so we decided to look at the problem together."

"Nobody thought we'd get down to light waves," Townes acknowledged, "but I thought, gee, I really want to get to shorter waves, and how to do that?" He sat down at a desk and proceeded to write down all the numbers and concluded that it would in fact be possible using the same process that he had used with the maser, "but just pushing it further and further and on down to light waves." The process could operate the same way that the maser did. Energy could be generated from molecules and atoms by picking out certain ones with excess energy and allowing the waves to interact with them and drain the energy out of them. In masers the result was amplified microwaves; in this device the result would be amplified light. Light waves were also different from microwaves in that they were much shorter to begin with—about twenty thousandths of an inch. By contrast, microwaves are about one hundredth of an inch or a millimeter in length.

But how to create a device to generate these intensified light waves? That was where Schawlow came in. His idea was to arrange a set of mirrors, positioning each one at the end of a cavity, then bouncing the light back and forth. In this way it would be possible to eliminate the amplification of any beams bouncing in other directions, in effect ensuring that the light would have only a single frequency. Schawlow and Townes were both excited by the possibilities, and in the fall of 1957, they began working out the principles of a device that could emit high-intensity light beams.

Townes called his idea for a new device an optical maser. The term "laser"—which stands for light amplification by stimulated emission of radiation—was apparently first used by a fellow researcher at Columbia named Gordon Gould.† Simply stated, a laser is a device that creates and amplifies a narrow, intense beam of coherent light. Atoms emit radiation; when they are excited, neon atoms in a neon sign, for instance, emit light. Normally, they radiate their light in random directions at random times. The result is incoherent light—a technical term for photons (the smallest unit of light) hurtling in every direction. What Townes and Schawlow wanted to do was create a device that would generate *coherent* light that would be synchronized at a single frequency and that would travel in a precise direction. To do this, though, they had to find the right atoms, and create an environ-

† Inspired by the maser, some of Townes's graduate students suggested an imaginary device for amplifying infrared waves: an "iraser." For gamma rays they came up with a "gaser" and for radio waves a "raser."

ment in which the atoms all cooperated, meaning that they would give up their light at the right time and move in the same direction.

Atoms and molecules exist at two energy states: high and low. Those atoms at low levels can be excited to higher levels, usually by heat. They give off light when they return to a lower level from a higher one. Ordinary light sources—a lamp, for instance—operate in such a way that the many excited atoms or molecules emit light independently of one another as well as in many different colors (wavelengths). That is, ordinary light is incoherent.

To produce coherent light, the electrons in the atoms of a laser medium (gas, for example) are first pumped or energized to an excited state by an external energy source. The external energy is in the form of photons (packets of light). Excited by these external photons, the photons within the laser chamber emit energy—a process called stimulated emission. The photons emitted from the laser travel in tandem with the stimulating photons—they are operating at the same frequency. As the photons move back and forth in the chamber between two parallel, silvered mirrors, they trigger additional stimulated emissions. As these stimulated emissions multiply, the result is coherent light, which consists of a single frequency, or color. This intense, directional, monochromatic light finally exits the chamber through one of the mirrors, which is only partially silvered for that purpose.

Experiments using the maser had shown that a crystal, such as ruby or garnet, would be ideal for such a device. (Gases, liquids, and other substances can also be used.) The crystal rod, placed in a laser cavity, acts as the repository for the atoms to be excited, so that more of them are at higher energy levels than are at lower energy levels. In Schawlow's scheme, the reflective surfaces at both ends of the cavity permit energy to reflect back and forth, building up in each passage. Photons, released by the excited atoms, reflect back and forth, in the process stimulating additional atoms to emit radiation. The emitted light is coherent, moving on the same frequency, and monochromatic, of a very pure wavelength, or color. When the photons build up enough energy, they surge through the partially silvered end of the ruby rod as a brilliant, powerful red beam of laser light.

However, before such a device could be realized, there were several technical questions that had to be overcome first, not the least of which was somehow getting around the second law of thermodynamics, which, in effect, told Townes that molecules cannot generate more than a certain amount of energy. The second law is based on the concept of entropy, or the measure of how close a system is to equilibrium. It can also be thought of as a measure of the disorder in the system. The law states that the entropy—that is, the disorder—of an isolated system can never decrease. Anytime an isolated or closed system achieves a state of maximum entropy, it can no longer undergo change: it has reached equilibrium (albeit a chaotic one). The second law poses an additional condition on thermodynamic processes. It is not enough to conserve energy and thus obey the first law. Energy has a price. In a closed system, in which no change is possible, where equilibrium has been attained, it would not be possible to produce energy from nothing. Energy could not be continually drawn from a cold environment to do work in a hot environment at no cost. That would be a perpetual motion machine, which keeps working even if no energy is ever consumed—and perpetual motion machines do not exist. They do not obey the second law; energy cannot be drawn from a cold environment to a hot one unless work is done, for example. Bouncing light back and forth inside a cavity in an attempt to generate higher states of energy seemed to Townes suspiciously similar to the idea of a perpetual motion machine.

The idea of how to get around the second law came to Townes while he was attending a scientific committee meeting in Washington, D.C., in 1957. Townes left his hotel room, which he was sharing with Schawlow, to go for an early morning stroll in nearby Franklin Park. As he walked he mulled over his problem. An idea occurred to him. He thought, "Now wait a minute! The Second Law of Thermodynamics assumes *thermal* equilibrium. We don't have to have that!" He was working with light, not heat.

He pulled an envelope out of his jacket and excitedly began to jot down a calculation, trying to determine how many molecules he needed in a resonator to get the power output he wanted. "And I suddenly saw where we could pick out molecules with a lot of energy and use these selected molecules, then send them into a cavity where the rays would bounce back and forth and take the energy out of the molecules."

Townes was certain he was onto something. "And I was just absolutely elated. It was just, 'Hey, this looks like it might work!' " He rushed back to his hotel room and showed his calculations to Schawlow. "I told him about it and he said, 'Okay, well maybe.' And so that's how the idea started, the whole thing."

"There's a sense of wonder," Townes said. "It's like a sudden revelation, an inspiration. . . . On the one hand it seemed almost too good to be true. . . . I thought the principles said it would work. I wasn't sure whether we could really make it work. . . . There were questions, but as well a great excitement, 'What the hay, this looks like a way of doing it.' "

The problem for scientists, Townes acknowledged, was that while they may know a lot, they can also get so set in their ways that they fail to notice what's lying off the beaten track. "When the maser came along, and the laser," he said, "it was sort of like opening a door—I didn't even know the door was there—and going to a room, a fantastic room that I never dreamed might be there at all." The achievement of Townes was to gather the pieces and put them together in the right way.

It is at this point in the history of the invention of the laser that things become murky. At the time of his discovery, Townes wondered why no one had discovered a laser (at least the idea for one) before. It wasn't as though he had come up with a brand-new idea, he acknowledged—all the principles had long been known. But Townes might have spoken a little too soon, or perhaps been just a bit disingenuous. In fact, Townes and Schawlow weren't the only ones involved in the research that gave the world the maser and the laser. And it's even possible that Townes wasn't the first to invent the laser.

Three other scientists—Joe Weber, a professor at the University of Maryland, and two Soviet scientists, Nikolai Basov and Aleksandr Prokhorov, in Moscow—were also making substantial contributions at the time Townes and Schawlow were trying to develop the laser. Weber, Basov, and Prokhorov all worked in the field of microwaves and molecules. In fact, Weber deserves much of the credit for discovering the basic principle for both lasers and masers. He had given the first public talk ever on the subject as early as 1952, and he was the first to realize it might be possible to detect gravitational waves—made up of minute distortions of space—predicted by Einstein's general theory of relativity. Weber was also the first to build the first detectors to search for these elusive waves. Townes himself acknowledged that Weber had been "one of the four people who really saw the idea" before anyone else had. It was just that Weber never received the recognition accorded Townes and others, partly because his innovative ideas often outpaced the technical tools available to realize them.

It is, however, Gordon Gould (b. 1920), coiner of the term "laser," whose claim to have invented the laser is, if anything, even stronger than Townes's. Born in New York City, Gould decided even as a child that he wanted to be an inventor like his idol Thomas Edison. As a graduate student at Yale, he specialized in spectroscopy and after obtaining an M.S. in physics in 1943, he went to work for the Manhattan Project, the top-secret program to build an atomic bomb. When the war was over, Gould continued his graduate studies in optical and microwave spectroscopy physics at Columbia University. In 1957, three years after Townes had demonstrated the first maser, Gould, then 37, became intrigued by the idea of developing the same kind of device.

In another one of those coincidences in which the history of inventions abounds, Gould has said that his first ideas for the laser "came in a flash" one night in 1957. Light waves reflected between two mirrors inside a gas-filled chamber, Gould realized, could be made to form a single concentrated beam that he estimated would superheat a substance in a millisecond. Even at this conceptual stage, Gould recognized the potential of the laser for industry, communications, and the military. Over the next three days, he proceeded to put his ideas down in a notebook that he titled "Some rough calculations on the feasibility of a LASER: Light Amplification by Stimulated Emission of Radiation." It is widely thought that this was the first use of the famous acronym. He then took the notebook to a candy store near his Bronx apartment to have it notarized. Had he gotten good legal advice at the time, he would have saved himself a good deal of trouble and money by applying for a patent as well—before Townes and Schawlow. But Gould held off in the belief that he would be better off building an actual prototype to prove that the idea would work in practice.

Abandoning his academic career, Gould teamed up with a small technology company called TRG. TRG used his idea for a laser as the basis for a proposal it submitted to the Defense Department. It was only then that Gould applied for a patent, months after Townes and Schawlow had applied for their own. Moreover, the two collaborators had already become identified with the invention of the laser on the basis of a paper titled "Infrared and Optical Masers," which was published in the December 1958 *Physical Review*, the journal of the American Physical Society. In addition, Townes had several advantages that Gould lacked, including important academic and government appointments and a vaunted reputation in scientific circles.

There still remained the question of whether a real laser could be made to work. And that wasn't the only thing that some found troubling. As Townes put it, "Many people thought, 'Well, okay, it's kind of an interesting idea, but where is it going? It isn't going to do anything.' " They considered it a solution in search of a problem. Townes found their attitude puzzling. "Look, it (the laser) marries optics and electronics. Both of those are very important. You marry those two fields and there are going to be lots of applications."

Confirmation of Townes's view would have to wait until an actual laser was up and running. The wait wasn't long. With the publication of Townes and Schawlow's paper, a mad scramble began, with scientists from academia and industry contending to be first to develop a working laser. The race went to an American physicist named Theodore Maiman at Hughes Laboratories, who constructed the first practical laser in 1960. Maiman had also been responsible for innovations in the design of the maser that had greatly increased its practicability. In 1960, three scientists at

Bell Labs constructed the first laser using a gas medium instead of a crystal. In the same year, Townes and Schawlow received a patent for their device—No. 2,929,922.

In 1964, Townes shared the Nobel Prize for physics with Prokhorov and Basov of the Lebedev Institute in Moscow for "fundamental work in the field of quantum electronics which has led to the construction of oscillators and amplifiers based on the maser-laser principle." (For his part, Schawlow shared the 1981 Nobel Prize for pioneering laser spectroscopy, not for inventing the laser.)

Ironically, the Nobel for his groundbreaking work on the maser and laser came after Townes was beginning to lose interest in investigating the potential of his own inventions. "It was an interesting field and I worked at it for a few more years," Townes said, "but then I decided, well, there are plenty of people working on this. There are other things to be discovered that are being neglected, and so I went on to other things." Townes felt that ten or fifteen years working on the same thing were quite enough.

In spite of the recognition Townes was accorded by the Nobel Committee, controversy over the discovery of the laser did not fade away. Initially, the patent dispute went against Gould. The case hinged on both the timing of Gould's application and the question of just how "diligently" he had developed his concept. The patent examiner determined that Gould's record did not, in fact, properly represent the invention, because his proof of prior conception was incomplete. The scientific community sided with Townes, regarding Gould as an "outsider," and scoffing at his claims. Even in the face of reversals by the Patent Office and the courts, Gould persisted in his crusade for vindication. To subsidize his legal costs, Gould traded a half interest in his pending laser patent to the Refac Technology Development Corp. based in New York. Then in October 1977, in a ruling that caught the laser industry by surprise, the Patent Office issued Gould a patent on one type of laser called an optical pumping laser. Refac, in turn, notified manufacturers of these lasers that henceforth they would have to pay royalties on their optical pumping lasers.‡

Even after the Patent Office decisions, several manufacturers of laser equipment refused to pay royalties to Refac. The legal battle raged on until 1987, when the court came down decisively in Gould's—and Refac's—favor, thirty years after Gould had his notebook notarized in a Bronx candy store. Gould now holds the basic patents covering optically pumped and discharge-excited laser amplifiers, which are used in 80 percent of the industrial, commercial, and medical laser applications. For all the expense and tribulation that the legal conflict caused Gould, he ended up making far more money from his share of the royalties than he would have made had his first patent application been promptly approved.

Gould's rival, Charles Townes, had meanwhile moved onto other interests. After a stint teaching physics at MIT and serving as a scientific adviser to the Johnson administration, Townes moved west in 1967 to become a professor of physics at the University of California at Berkeley. There he returned to one of his first loves—radio astronomy—which he was forced to abandon at Bell Labs when the exigencies of wartime science took precedence. In their radio probes of interstellar space, Townes and his associates discovered water and his old favorite, ammonia, the first complex molecules found outside our solar system. That ammonia molecules in space, as on Earth, should figure so prominently in Townes's achievements (first with the maser, then with radio astronomy) seems to intimate at the hand of some curious fate at work. Townes has told interviewers:

‡ The following year, Gould received a second patent on a broad range of laser applications.

> I have tried to devote my energies to opening new fields with the intention of moving on to other things as soon as they start to become well established. I like to turn over new stones to see what is under them. It is the most fun for me to be on the fringes, exploring aspects that seem interesting to me but have not attracted the attention of others. Once a field is opened up and is successful, and once others are flocking to it, I feel my own efforts in it are no longer critical, and about that point I like to move on to something I think may be promising but overlooked.

However the Patent Office or the courts settled the controversy over the origin of the laser, it is still possible to argue that both Townes and Gould came up with the idea for the laser—that, in a sense, the idea was out there, and that they both had the luck, the intuition, and the knowledge to reach out and grab hold of it almost simultaneously.

CHAPTER 9

The Pioneer of Pangaea

Alfred Wegener and the Theory of Continental Drift

The hypothesis, critics said, was daring . . . spectacular . . . imaginative . . . and wrong. It was, in the words of one skeptic, a hypothesis of a "footloose type in that it takes considerable liberty with our globe, and is less bound by restrictions or tied down by awkward, ugly facts than most of its rival theories." More shocking still, the critic went on, it "leaves the impression that it has been written by an advocate rather than by an impartial investigator." What's more, the detractors agreed, the hypothesis was "not scientific, but takes . . . a selective search through the literature for corroborative evidence, ignoring most of the facts that are opposed to the idea, and ending in a state of auto-intoxication in which the subjective idea comes to be considered—objective fact." Even experts who were willing to give the hypothesis the benefit of the doubt suffered misgivings. The hypothesis, wrote one, "shows little respect for time-honored ideas backed by weighty authority. Its daring and spectacular character appeals to the imagination both of the layman and of the scientist. But an idea that concerns so closely the most fundamental principles of our science must have a sounder basis than imaginative appeal."

And what was this hypothesis that could arouse such fury and vituperation? It was called continental drift, and what it proposed was the curious idea that over the course of many millions of years the shape of the Earth's surface had undergone a dramatic transformation as continents moved, mountains rose and fell, and oceans invaded territory once claimed by land.

Rancorous debates over how the Earth was formed and whether it was still in the process of being reshaped were nothing new in the history of geology. The acceptance of a new theory and the debunking of a previously accepted one could ruin careers and even cast into doubt deeply held religious beliefs. To understand the context of the debate, it's necessary to go back to the beginning—at least metaphorically speaking—to the Book of Genesis. Throughout much of the seventeenth and eighteenth centuries, geology was still dominated by the concept of the biblical Flood as a major force in the formation of the Earth's surface. Long-ago catastrophes, it was widely held, had brought about sudden and radical changes that persisted to the present day. In 1756 a German theologian named Theodor Christoph Lilienthal, in observing the similarity of certain coastlines—that of South America and Africa, in particular—drew on biblical references to conclude that the Earth's surface had been torn apart in the aftermath of the Flood. The same phenomenon caught the attention of Francis Bacon (1561–1626). Bacon, one of the foremost essayists of the seventeenth century and lord chancellor to James I, also pointed out the similarity in shape between Africa and South America. However, he failed to make much out of the fit of their opposing coastlines.

Then early in the nineteenth century, the German naturalist and explorer Alexander von Humboldt (1769–1859) proposed that the Atlantic had been carved out by catastrophic water action to form a gigantic "valley" with parallel sides. In a spirited attempt to explain the similarity of coastlines on either side of the Atlantic, another seventeenth-century naturalist, Antonio Snider-Pellegrini, suggested that the Flood had thrown up a great outpouring of material from within the Earth that then separated the landmass and pushed the continents apart, with formation of the Atlantic as the result. To support his thesis, Snider-Pellegrini also cited the remarkable similarities of fossils and certain rock formations on both sides of the Atlantic, going so far as to produce a map that reassembled the Americas, Europe, and Africa like pieces of a jigsaw puzzle.

In an even more adventurous theory, the Moon was wrested down from the sky (at least on paper) to account for both the Atlantic and the Pacific. According to this theory, first proposed by George Darwin (1845–1912), a mathematician and an astronomer (and the son of Charles), the Moon had been thrown off the Earth long ago when our planet's spin was far faster than it is today. The violence caused by jettisoning the Moon, Darwin contended, had left a great cavity, which was subsequently occupied by the Pacific, while at the same time the flow of material needed to fill this cavity dragged the Americas away from Europe and Africa.

Other scientists looked to the interior of the Earth to account for what could be observed on its surface. The nineteenth-century English chemist Humphry Davy, who was the first to isolate a number of the chemical elements, proposed that there were continuous "fires" within the interior of Earth, essentially an unceasing process of oxidation that provided a sustainable source of heat. It was a theory that he later repudiated as implausible. What was causing these fires to burn in perpetuity? No answer suggested itself.

Davy's dilemma was resolved by the discovery of radioactivity at the end of the nineteenth century. It turns out that radioactive elements such as uranium, thorium, and a radioactive form of potassium are found in many kinds of rock; their slow decay into other elements generates heat. While this heat dissipates so rapidly when it reaches the surface of the Earth that it is virtually unobservable, it accumulates deep inside the Earth, where it produces a continual source of energy.

In the middle of the nineteenth century, William Thomson, Lord Kelvin calculated that, even if the Earth at its birth had been as hot as the surface of the Sun, it could not be more than a few hundred million years old and still have left as much internal heat as it in fact did.* This came as bad news to geologists. Their study of rock formations told them that the Earth needed to have been around for eons longer in order to account for the complex geological features they observed. That great mountain chains had risen and then crumbled had been well established, but mountains required far more time than a few hundred million years to reach their lofty heights and shrink to unprepossessing hills.

By the early twentieth century, the allure of past catastrophes, biblical and otherwise, to explain the current state of the Earth began to fade, giving way to a new theory, with the tongue-twisting name of "uniformitarianism." In this view, proposed by the nineteenth-century geologist Charles Lyell, all the forces that can be seen at work on the Earth today, operating over long periods of time, should be adequate to explain what has happened in the past.

---

* Lord Kelvin derived the measurement of absolute temperature that bears his name today.

But as more scientific evidence in the form of fossils became available, the theory of uniformitarianism began to be called into question. Geologists discovered, for instance, that the polar ice caps could have rivaled Caribbean isles in climate and tropical flora and fauna. In 1908, an American scientist, Frank Bursley Taylor, came up with an idea that turned George Darwin's on its ear. In Taylor's theory, the Moon came so close to Earth during the Cretaceous period, 100 million years ago, that it was captured by our planet's gravity. The resulting tidal pull dragged the continents toward the equator. In the process, the great mountain chains of the Himalayas and the Alps were created. Looking heavenward to account for earthbound phenomena was all the rage. Another American, Howard B. Baker, proposed that a close approach of Venus hundreds of millions of years ago pulled enough rock out of the Earth to form the Moon and at the same time set the continents in motion. By the dawn of the twentieth century, a new theory arose that held that land bridges, now sunken, had once connected far-flung continents.

However, it fell to a German meteorologist and explorer named Alfred Lothar Wegener (1880–1930) to amass all the evidence to overturn orthodox science of the day, and in doing so cast out the need for biblical floods, Venus, or the Moon to explain what actually was happening to the Earth's crust. His particular genius lay in the way that he tapped several diverse fields to obtain his evidence—something rare in the scientific practice of his day—drawing on meteorology, geology, oceanography, seismology, geomagnetism, paleontology, evolution, even mountaineering and polar exploration.

Alfred Wegener was one of that breed of scientist-explorers who dominated polar exploration early in the twentieth century. One of his associates described him as "a man of medium height, slim and wiry, with a face more often serious than smiling, whose most notable features were the forehead and the stern mouth under a powerful, straight nose." Another offered a somewhat different characterization, calling him a "quiet man with a charming smile." The son of a minister, Wegener was born in Berlin in 1880 and from an early age displayed an athleticism that would serve him well in his future explorations. He initially set out to study astronomy, but no sooner had he obtained his Ph.D. than he dropped astronomy and turned his attention to meteorology, which at the turn of the century was still a new science. Not content with studying the weather from the ground, he decided to carry out his observations in the air. This was the heyday of ballooning, and in April 1906, Wegener and his brother Kurt broke the world's record for long-duration flight in a free balloon—fifty-two hours across Germany and Denmark, out across the Kattegat (the body of water separating Denmark and Sweden), and back over Germany. That same year, the 26-year-old Wegener was invited to join a Danish expedition setting out for Greenland, a country that had long obsessed him. It was an obsession that would one day prove fatal.

After two years in Greenland conducting meteorological observations, Wegener returned to Germany to become a lecturer in astronomy and meteorology at the University of Marburg, and it was there he established what was to be a lifelong friendship and professional association with a like-minded student, Johannes Georgi. Georgi later agreed to work under Wegener when the latter was appointed head of the Meteorological Research Department of the German Marine Observatory in Hamburg.

One day in the autumn of 1911, while browsing in the library at the University of Marburg, Wegener, then 31, happened to come across a scientific paper that listed fossils of identical plants and animals found on opposite sides of the Atlantic. Like others before him, Wegener was taken by the striking fits of some opposing coastlines, notably the way in which the bulge of Brazil conforms to the curve in the coast, or bight, of Africa on the opposite side of the Atlantic. "It is just as if we were to refit the torn pieces of a newspaper by matching their edges and then check whether the lines of print run smoothly across," he wrote. "If they do, there is nothing left but to conclude that the pieces were in fact joined in this way."

The apparent match-up of coastlines was just the beginning. The more he looked, the more evidence he found for a link between the various continents separated by vast bodies of water. The geological composition of the Appalachian Mountains of eastern North America matched that of the Scottish Highlands, while the distinctive rock strata of the Karoo system of South Africa were identical to those of the Santa Catarina system in Brazil. Abundant evidence also existed to indicate that great climatic changes must have occurred in the past, suggesting that landmasses, whose climates are now dramatically different, might once have been joined together. The Arctic island of Spitsbergen, for example, had yielded fossils of tropical plants such as ferns and cycads, and discoveries of coal in Antarctica showed that that region had once been tropical. In South Africa, on the other hand, deposits of mixed sand, gravel, boulders, and clay—typically left behind by a melting ice sheet—showed that South Africa once had a much colder climate than in Wegener's day. The large gypsum deposits, formed in Iowa, Texas, and Kansas during the Permian period 250 million years ago, pointed to a hot, arid climate in those regions; the same conclusion could be drawn from the finding of salt deposits in locations as far apart as Kansas and Europe.

Wegener also learned of uncannily close relationships of species now widely separated. It turned out, for example, that lemurs, the most primitive of the primates, are found only in East Africa, on the island of Madagascar, and across the Indian Ocean. By the same token, the hippopotamus is found on Madagascar and in Africa. Assuming that the animals evolved on the mainland, how could they have swum 250 miles of open sea to reach Madagascar?

The contention by the geologists of his day that these many similarities were due to land bridges long since vanished didn't seem reasonable to him. Then how *could* all these similarities be explained? Wegener realized that he now had all the pieces to solve the puzzle. The fossil evidence was one piece, the geological evidence another, the species evidence still another, and the climatic shifts still another. It was as if Wegener were viewing the problem from above, aloft in his balloon, taking in all aspects of the landscape.

Suddenly he thought he had the answer. All these pieces represented a part of the same phenomenon. Suppose, he thought, that the continents had once been joined together and that gradually, over hundreds of millions of years, had somehow split apart and moved about the surface of the Earth, eventually reaching their present locations?

The idea seemed preposterous. It was also true.

It was in a talk that he presented to the Frankfurt Geological Association in January 1912 that he proposed the idea of what he termed "continental displacement," or what later became known as continental drift. Though this idea was greeted with skepticism, Wegener was not deterred. "A conviction of the fundamental soundness of the idea took root in my mind," he wrote.

Unfortunately, Wegener's plan to pursue his research and gather corroborative evidence to support his thesis had to be postponed because of World War I.

As a young reserve lieutenant in the German army, Wegener was twice wounded—first in the arm and later in the neck. As he lay recuperating in the hospital, he began to ponder the ideas he had raised two years earlier. With little to do while he recuperated, he continued to pursue his research, which culminated in 1915 with the publication of the first comprehensively developed theory of continental drift, *The Origin of Continents and Oceans*. (Expanded editions were published in 1920, 1922, and 1929.) He declared in *Origin*:

> Scientists still do not appear to understand sufficiently that all earth sciences must contribute evidence toward unveiling the state of our planet in earlier times, and that the truth of the matter can only be reached by combing all this evidence. It is only by combing the information furnished by all the earth sciences that we can hope to determine "truth" here, that is to say, to find the picture that sets out all the known facts in the best arrangement and that therefore has the highest degree of probability. Further, we have to be prepared always for the possibility that each new discovery, no matter what science furnishes it, may modify the conclusions we draw.

About 300 million years ago, until the Carboniferous (or Coal-Forming) period, Wegener maintained, the continents had formed a single supercontinent, called Pangaea (from the Greek for "all the Earth"). Pangaea had rifted, or split, and its pieces had been moving away from each other ever since. He even prepared maps showing how each of the present-day continents had once fitted together. The breakup occurred in successive ruptures: Antarctica, Australia, India, and Africa began separating in the Jurassic period—the age of dinosaurs—some 150 million years ago. In the subsequent Cretaceous period, Africa and South America separated "like pieces of a cracked ice floe." The final separation divided Scandinavia, Greenland, and Canada at the start of the ice ages, about a million years ago.

Wegener further argued that the Mid-Atlantic Ridge that forms such islands as Iceland and the Azores was composed of continental material left behind when the continents that now flank the Atlantic broke apart. He proposed that a land called Lemuria once linked India, Madagascar, and Africa; if this were so, then it would explain the widespread distribution of the lemur and the hippopotamus. (Madagascar is now believed to have separated from the African mainland about 165 million years ago.) That marsupials such as the kangaroo and the opossum are found only in Australia and the Americas impelled Wegener to link Australia with distant South America. The distribution of fossil remains of trees and other plants that flourished during the Carboniferous period provided further fodder for Wegener's thesis. These fossils were typified by a seed fern, known (because of its tonguelike leaves) as *Glossopteris*. Each region of the Earth, in the past as well as now, tends to have its characteristic vegetation. Yet botanists have found that the *Glossopteris* vegetation thrived in such widely separated regions as India, Australia, South America, and South Africa. Remains of this vegetation were even found embedded in coal seams in mountains near the South Pole by members of the Scott and Shackleton expeditions.

The continents are like great barges, Wegener said. A loaded barge will sink while the water beneath it flows aside to make room for the barge's greater submerged volume. When the barge is unloaded, the water flows back in to lift the barge to a height commensurate with its reduced weight and greater buoyancy. Wegener believed that the ocean floors represented a deeper layer of the Earth, formed of material in which the continents are "floating." The material of such a layer would have to be denser than continental rocks, just as water is denser than the ice of an iceberg. If this were not the case, the iceberg would not float.

What this meant was that, in effect, the ocean floor is made up of different stuff than the continents. If there had been any intercontinental land bridges that had vanished, they should now be part of the ocean floor. But if it could be proven that the ocean floor had an entirely different composition, Wegener's case would be that much stronger. However, the methods necessary to measure the gravity and density of the ocean floor, or to dredge up specimens from far enough down, were beyond the technology of the day to definitively establish his hypothesis. Wegener acknowledged as much. "It will be impossible for a long time yet to bring up samples of rock outcrop from these depths either by dragnet or other means," he wrote.

Then Wegener turned his attention to the shifting of the polar axis over long periods of time. Early in the twentieth century, the Hungarian physicist Baron Roland von Eötvös (1849–1919) had calculated that, because the Earth is spinning and its shape bulges at the equator, there should be a very slight force nudging blocks of material in the direction of the equator. This force, which Wegener called the "pole-fleeing force," together with a tendency for tidal drag, might account for continental movements. If the spin axis of the planet changed from time to time, causing the pliant Earth to adjust its shape to produce a new equatorial bulge and to alter the climate of different regions, it might also be responsible for new directions of drift.

The theory of continental drift also appeared to resolve an old dilemma about how mountains rose and eroded. This view of how mountains are created flew in the face of a traditional idea that they had formed by the shrinkage of the entire planet, a theory that was favored by Isaac Newton. In 1681 Newton had proposed that mountains had erupted from the crust of the Earth while it was still being formed, before "the settling and shrinking of the whole globe after the upper regions or surface began to be hard." But by Wegener's time, geologists had shown that mountains continued to be built long after the earliest stages of crust formation. According to the traditional view, the interior of the Earth had once been an inferno, but over time had gradually cooled. This cooling process accounted for the shrinkage of the Earth's surface. As geological evidence indicates, though, the Earth hasn't cooled appreciably, an argument against the simple shrinkage of the world. Wegener contended that it was far more likely that mountain building derived from the pressures of continental drift than from any overall shrinking.

According to Wegener, the mountains that line the western coasts of the Americas, from the Andes to Alaska, were wrinkles of the Earth's crust, which formed by pressure as the great blocks of continental rock pushed west. The mountains of New Guinea were similarly pushed up by the northward drive of the Australia–New Guinea block of continental material, while the mountain ranges of Asia, from the Himalayas north to the Tien Shan on the borders of the former Soviet republics of Central Asia, were produced by the northward pressure of the subcontinent of India.

Continental drift, Wegener said, is still going on. Subsequent research has borne him out. The northern Baltic is rising about one meter every century. The north shore of Lake Superior has been rising steadily and a tide gauge on the pier at Fort Churchill, in Hudson Bay, shows an uplifting of about two meters per century.

Reaction to Wegener's theory was almost uniformly hostile. One of the major problems that critics seized upon was the absence of a convincing mechanism for how the continents might move. In Wegener's conception, the continents were moving through the Earth's crust, like icebreakers plowing through ice sheets, propelled by centrifugal and tidal forces. Impossible! railed Wegener's opponents. If that were the case, the shape of the continents would be distorted beyond recognition. Besides, the centrifugal and tidal forces were far too weak to move continents. One scientist calculated that a tidal force strong enough to move continents would cause the Earth to stop rotating in less than one year. Some of Wegener's original data were also flawed, which caused him to make extravagant predictions that were simply wrong—he suggested, for instance, that North America and Europe were moving apart at over 250 cm per year (about ten times the fastest rates seen today, and about a hundred times faster than the measured rate for North America and Europe). Detractors offered other objections: animal species on opposite sides of the Atlantic were not as similar as Wegener claimed, they said, and the geology didn't bear as much resemblance as he said, either. Wegener, wrote one, "generalizes too easily from other generations." The fiercest attack came from Edward W. Berry, a professor of paleontology at Johns Hopkins University, who declared that Wegener's methodology wasn't even scientific and that he had deliberately ignored any evidence contrary to his thesis. Even Chester R. Longwell of Yale, who was not entirely unsympathetic to Wegener, was inclined to express skepticism. "We know too little about the Earth and its history to indulge such final judgments," the geologist wrote. "Why must we—or how can we—make up our minds finally about the matter while we admit great gaps in critical geologic information and in knowledge of geophysical principles?" While he wasn't quite willing to write off the theory entirely, he added, "To me the concept of moving continents is *hypothesis*, and as such it must be a target exposed to merciless fire of fact-finding and critical analysis."

And merciless fire was what it got. Though a few scientists supported Wegener—the South African geologist Alexander Du Toit believed that it was a good explanation for the close similarity of strata and fossils between Africa and South America, for instance—the majority of geologists of Wegener's day continued to believe in static continents and land bridges. The attacks on continental drift reached their full fury in 1926 at a meeting of the American Association of Petroleum Geologists, where Wegener was denounced in such scathing terms that one writer was prompted to call it an "ambush" more than a scientific seminar. Even Wegener's own father-in-law voiced his opposition to the theory (probably because he was annoyed that Wegener had abandoned meteorology for the more uncertain science of geophysics). That so many of his detractors were established scientists only made his cause more difficult to defend.

Wegener remained undaunted. Nothing if not tenacious, he denounced his critics as "narrow-minded" and resolved to find more evidence that would establish the correctness of his theory. After the disastrous 1926 symposium, Wegener tried to strengthen his argument with further elaborations and revisions of his book. He produced what he considered the "first precise astronomical proof of a continental drift," based on observations that seemed to show that Greenland and the United States were drifting relative to Europe. But when the Danish team responsible for the original data, derived in 1906–1908, went back and checked again in 1936 and 1948—long after Wegener's death—

they discovered no evidence for drift at all. What they didn't realize at the time was that the expected rates of movement were much too slow to become evident through the ordinary position-determining methods that they were using.

While Wegener continued to concentrate on continental drift, he still needed to make a living. He was having great difficulty in securing an academic post, mainly because he failed to fit the traditional image of a research scientist. It was regrettable, his friend and former student, Johannes Georgi, wrote, "that this great scholar, predestined for research and teaching, could not get a regular professorship at one of the many universities and technical high schools in Germany. One heard time and again that he had been turned down for a certain chair because he was interested also, and perhaps to a greater degree, in matters that lay outside its terms of reference—as if such a man would not have been worthy of any chair in the wide realm of world science." His job search was finally rewarded in 1926 by a post as a professor of meteorology and geophysics at the University of Graz in Austria.

It was on the initiative of Georgi that Wegener decided in 1930 to undertake another expedition to Greenland. It would be Wegener's fourth trip to Greenland—and his last.

Throughout the late 1920s, Georgi had been making weather observations from a base on Iceland and had detected a very high and intense flow of air from the direction of Greenland—what is now known as the jet stream. He proposed setting up a winter station on the Greenland ice sheet to learn more about this flow, which clearly had an effect on determining the weather in Europe. Since he regarded Wegener as "the greatest expert on Greenland," Georgi naturally wanted him to come along. Wegener was reluctant. He liked the idea—it should have been done long ago, he said, but for the war—but complained that he was no longer young enough to endure such an arduous expedition. Nonetheless, such was his fascination for the Far North that he agreed to go anyway.

The main goal was to set up a station to carry out weather observations throughout the polar night. But plans went awry from the start. Ice and storms delayed unloading of supplies. After an agonizing succession of delays, Wegener set out from the coast on September 21, 1930, with a party of fourteen, almost all Greenlanders, and several sledges carrying four thousand pounds of supplies that would allow Georgi and a colleague already at the ice station to make it through the winter. The weather was punishing; after pushing on for a hundred miles, the driving snow and temperatures as low as 65 degrees below zero caused all but Wegener and two other members of the relief team to turn back. Wegener, however, was determined to make contact with the two men at the station and if necessary help them escape if they wished. There was no way, with the team reduced to three, that they could possibly transport all the supplies they had brought with them.

Wegener and his two companions succeeded in reaching the station, but the triumph was brief. With their dogs so worn out from the trek, Georgi noted in his diary, it would be "a race with death" to try to return to the coast. But there was no choice with barely enough food left to feed two men, let alone five. So it was agreed that three would stay on at the station, dividing rations as frugally as they could, while Wegener and a Greenlander named Willumsen would set out for the coast.

They never made it. Wegener's body wasn't found until the following summer, when conditions permitted a search. It appeared that he hadn't frozen to death but instead had succumbed to a heart attack. He died a few days after his fiftieth birthday.

Wegener's theory of continental drift went ignored for decades. This was one case where science stalled until technological advances gave it a needed shove. It was the intensified exploration of the Earth's crust—especially the ocean floor—beginning in the 1950s that largely accounted for the revival of interest in continental drift. By the late 1960s, a modified form of the theory of continental drift, known as plate tectonics, was embraced by almost all geologists, in sharp contrast to the ferocious reaction to it forty decades earlier. That it had taken geologists so long to come around is an indication of how fiercely they had clung to the belief that the continents and oceans have always occupied the same place. Of course, their surrender could only have occurred because the proof in favor of drift was so compelling.

Since Wegener's death, data collected from the ocean floor have confirmed Wegener's theory that its composition does in fact differ from that of the continents. Dredging devices have retrieved rock fragments that almost invariably are denser than rocks, like granite, ordinarily found on continental terrain. Additional support for the theory comes from sampling the entire sequence of sediment layers in the ocean floor, using sophisticated drilling technology. Experiments that would have been impossible in Wegener's time have also established that gravity in the oceans differs little from gravity on land. This finding suggests that the ocean floor is very dense, as Wegener had predicted, and thus is "floating" at its proper depth. What makes this finding so important is that it ruled out the possibility that continents, or land bridges of some kind, had sunk into the oceanic depths.

In one significant respect, however, Wegener was wrong and his critics were correct—at least to some degree. Continents, it turns out, do not plow through the ocean floor like icebreakers. Instead, both continents and the ocean floor form solid plates, which "float" together on the underlying mantle, called the asthenosphere. The mechanism of the movement, which had eluded Wegener, turns out to be the geothermal energy produced in the Earth's interior, although this process is not yet fully understood.

Because both the plates and the ocean crust move in tandem, the term "continental drift" is no longer considered quite accurate. Plate tectonics theory, which is its modern incarnation, holds that the Earth's crust is divided into a number of plates, which move horizontally at rates of a fraction of an inch to a few inches per year. New plate material is formed at their originating ends and old plate material is absorbed, or subducted, back into the Earth at their trailing ends. The new plate material consists of molten magma that has been disgorged from the depths of the ocean. Meanwhile the old plate material is carried back down into the mantle of the Earth, principally along the major earthquake zones surrounding the Pacific Ocean.

These plates are now being studied in order to better understand and predict earthquakes. In their exploration of the ocean floor, scientists have mapped and explored the great system of oceanic ridges—the sites of frequent earthquakes—where molten rock rises from below the crust and hardens into new crust. It has now been shown that the farther away you travel from a ridge, the older the crust and the older the sediments on top of the crust become. This is because the ridges demarcate the sites where plates are moving apart. On the other hand, where these tectonic plates collide, great mountain ranges may be pushed up. The Himalayas are one result of this process, as the plate carrying India slowly drives into the Asian plate. Alternatively, if one plate sinks below another, deep oceanic trenches and chains of volcanoes are formed. Because earthquakes occur most frequently along the boundaries of these plates, they provide a useful indication for the mapping of the plate boundaries and depths. It is even possible to measure the speed of continental plates with extreme accuracy, using satellite technology. In a related experiment, several laser reflectors were placed on the Moon by American astronauts (and installed aboard Soviet unmanned vehicles as well), which, by beaming intense beams of light back to Earth, are intended to measure whether the continents are currently in motion with respect to one another.

As is often the case, brilliant ideas may languish for years, or else be cast into disrepute, because the technology has not been invented, or the necessary scientific understanding achieved, to prove their validity. This was certainly the case with continental drift, without which modern plate tectonics theory could not have been developed. Had Wegener survived his ordeal in the icy wastes of Greenland and lived into his eighties, he would have seen his once reviled thesis almost entirely vindicated.

CHAPTER 10

Solving the Mystery of Mysteries

Charles Darwin and the Origin of Species

Few scientific theories have stirred such controversy and skepticism—even after the passage of more than a century—than the theory of evolution. The weight of scientific scholarship and empirical evidence has failed to silence detractors, many arguing from deeply held religious beliefs, who contend that the theory is riddled with flaws and in any case has never been satisfactorily proven. Skeptics assert that, while it may have the support of a majority of scientists, the theory doesn't necessarily mean that it is valid, or at the very least that it should be given no more credence than a competing theory such as creationism. The volatile debate over evolution has grown especially contentious in the United States in recent years, spilling over into elections for school boards and the selection of textbooks for high schoolers. Many scientists have reacted with growing alarm to the recent attacks on evolution, fearing that a generation of American children will go through school without ever getting the scientific grounding they need to compete in the twenty-first century.

So inevitably do people associate the concept of evolution with the name of Charles Darwin that it's understandable why they might assume that Darwin was the first to conceive of the idea. But in fact, the *idea* of evolution had been gaining currency, at least within intellectual circles, for many years before Darwin and the English naturalist Alfred Russel Wallace (1823–1913) proposed their theories of evolution independently in 1858. What Darwin did (as well as Wallace, though with considerably less public recognition) was to discover the *mechanism* for evolution, which previous scientists had failed to do.

The theory of evolution rests in large degree on the pioneering work done by geologists and paleontologists of the seventeenth and eighteenth centuries; their discovery that the Earth's history stretched back eons and that it harbored in bedrock, shale, and sediment the fossilized remains of species that no longer survived posed a challenge to the biblical version of how the world came into being. James Ussher (1581–1656), archbishop of Armagh, famously placed the date of Creation 3,760 years before the commencement of the Christian era, specifically the afternoon of October 23, 4004 B.C.

The idea that the Earth might be much older than anyone had previously imagined was as revolutionary as it was unsettling. Beginning in 1749, the French naturalist Georges-Louis Leclerc, Comte de Buffon (1707–1788), set out to write his *Natural History*, one of the earliest accounts of the global history of biology and geology that had nothing to do with the Bible; his was a gargantuan enterprise that went on for nearly forty years. He had no doubt that the Earth was very ancient, venturing an estimate of nearly two hundred thousand years old. Apparently, though, he was prepared to concede in private that the truth might be closer to half a million years. (In fact, the Earth is about four billion years old.)

Then in 1809, another French naturalist, Jean-Baptiste de Lamarck (1744–1829), published his *Philosophie zoölogique*, which contained the first detailed statement of the principle that the progressive complexity of life-forms, from simplest to most advanced, runs parallel to the chronological development of life-forms. However, his theory is based on the idea that any naturally occurring variations were a result of the effects of the environment and that these variations could then be instantly inherited. The variations that would endure and continue to be transmitted to subsequent generations were those that made the creature best suited for its basic needs and its ability to survive in a particular environment. Possibly the best-known example of Lamarck's theory is his contention that giraffes first developed their long necks in order to browse for food on the high branches of trees. Subsequently scientists found no evidence that inheritance functions in this manner and Lamarck's theory was discredited.

The accelerated pace of geological and paleontological exploration (especially on shells) in the eighteenth century spurred even more adventurous theories. *Principles of Geology,* written by the influential British geologist, Charles Lyell (1797–1875) and published in 1831, dealt a further blow to the biblical account of Creation, establishing a scientific principle of the uniformity of nature free of supernatural intervention. Lyell was convinced that any changes in both the physical structure of the Earth and its inhabitants were governed by the same laws of nature that had been operational when the world first came into being. There was, he asserted, a uniformity of causes and effects. But he did not believe that the geological evidence supported any theory "of the successive development of animal and vegetable life, and their progressive advancement to a more perfect state." He was, however, willing to allow for the competition of species for food and territory, which left some winners and some losers. In this "universal struggle for existence," he wrote, "the right of the strongest eventually prevails, and the strength and durability of the race depends mainly on its proliferations."

That Charles Darwin should have become so interested in evolution—even though he never set out to become a scientist—may not be so surprising given the intellectual influence of his own grandfather, Erasmus Darwin (1731–1802). A significant figure in late-nineteenth-century natural history, a radical and a freethinker, the elder Darwin concluded that existing life-forms must have evolved gradually from earlier species. In his *Zoönomia* (1794–1796), a popular book on animal life, he ascribed evolutionary development to the organism's conscious adaptation to its needs and to the environment. In another book, *The Botanic Garden* (1794–1795)—something of a curiosity since it is all in verse—Erasmus wrote, "All vegetables and animals now existing were originally derived from the smallest microscopic ones, formed by spontaneous vitality." His views on the subject echo Lamarck's idea that natural variation is principally caused by environmental influences. What was still needed was a theory that would resolve these conflicting views and offer a convincing explanation as to how life-forms evolved in time and complexity.

Charles Robert Darwin, the fifth child and second son of a prosperous country doctor, Robert Waring Darwin, and his wife, Susannah, was born in the English town of Shrewsbury on February 12, 1809. The family was comfortably upper middle class and there was never any time in his life that Darwin needed to worry about money. From all accounts, he was especially close to his mother, and there is at least some anecdotal evidence that she was responsible for inspiring him with a love of nature. (Her father, the potter Josiah Wedgwood, was a keen amateur naturalist.)

In 1817, when Charles was barely past his eighth birthday, his beloved mother died, leaving Charles in the care of his busy father and three older sisters. A year later Dr. Darwin packed the boy off to the prestigious Shrewsbury School, but Charles showed so little interest in his studies that his frustrated father declared, "You care for nothing but shooting, dogs and rat catching, and you will be a disgrace to yourself and to all your family." The problem was that while Charles was interested in chemistry and natural history, the school's curriculum, which emphasized the classics, simply bored him. In spite of his indifferent academic performance, Dr. Darwin still nurtured the hope that his son might become a doctor and enrolled him at age 16 in the famous medical school in Edinburgh. But as soon as Darwin witnessed his first operation, he was so repulsed he instantly gave up any idea of a future in medicine. Nonetheless, his stay in Scotland proved fortuitous since it led to a friendship with the zoologist Robert Grant and the geologist Robert Jameson. It was Grant who introduced him to the study of marine animals, and it was Jameson who encouraged his interest in studying the history of the Earth. Their influence didn't rub off immediately, though.

Having defied his father's wishes, Darwin soon drifted back into the easygoing lifestyle of a country gentleman. Finally he submitted to his father's demand that he get some kind of formal training to prepare himself for a career, and at the start of the academic year, 1828, he entered Christ's College, Cambridge, as a divinity student. He took his studies at Cambridge no more seriously than he had at Shrewsbury and Edinburgh. He fell in with a group of students who treated the university as a kind of sportsmen's club. Little by little the prospect of being ordained as a Church of England minister—his ostensible purpose in coming to Cambridge—lost all its allure. Yet, in a way that his father could not possibly have envisioned, Cambridge turned out to be an advantageous choice for Charles, after all. At the time there were a number of distinguished scientists at Cambridge, who counted as their leader a cleric-botanist by the name of John Stevens Henslow (1796–1861). Darwin soon became a regular at Henslow's "open houses" and was so often seen accompanying Henslow on his daily walks that he became known as "the man who walks with Henslow." It was no wonder that Darwin sought out his company; Henslow not only encouraged the young man's enthusiasm for science, but gave him the confidence in his own scientific abilities. In the early nineteenth century there were basically two accepted channels of gaining entry into the scientific world: mathematics (which included the physical sciences) and medicine (which included physiology). Darwin had mastered neither. To be taken so seriously by a scientist like Henslow must have seemed like an extraordinary validation.

In the spring of 1831, as Darwin was preparing to leave Cambridge, Henslow recommended that he accompany the geology professor Adam Sedgwick (1785–1873), on a three-week tour of North Wales to learn geological fieldwork. While Darwin had no liking for geology, he nonetheless could see where an expedition might be of some benefit. Just before he was to set out, he received the first volume of Charles Lyell's major work, *Principles of Geology,* as a gift. In studying Lyell's work, Darwin effectively taught himself the subject he loathed. And in the process, he learned to love it.

In the first volume of *Principles,* Lyell argued that the face of the Earth had changed gradually over long periods of time through the continuing, cumulative effects of various disruptions, such as volcanic eruptions, earthquakes, and erosion. These disturbances, Lyell said, had existed in the distant past and could be observed in the present. His view was dramatically at odds from the one held by most contemporary geologists, who maintained that significant geological changes resulted from short-lived events of great violence that could raise mountains or flood the entire planet. Lyell's work was to have considerable influence on Darwin's thinking.

Henslow continued to retain an interest in Darwin's future. In August 1831, he recommended Darwin for a position as unpaid naturalist on the HMS *Beagle*. The expedition had as its primary purpose the surveying of the east and west coasts of South America. Henslow felt that Darwin would not only prove an acute observer but would also make a more suitable companion for the aristocratic young captain, Robert FitzRoy, than the ship's regular naturalist-surgeon, whom FitzRoy considered his social inferior. Darwin's father at first refused permission on grounds that the voyage was dangerous and in any case wouldn't advance Charles's career. It took the intercession of Charles's brother-in-law to change his mind. In a letter to FitzRoy accepting the post, Darwin declared that he expected the voyage to be a "second birth." On December 27, 1831, the *Beagle*, a ten-gun brig that had been refitted as a three-masted bark, set sail from Plymouth, England. The voyage, planned for two years, ended up lasting five.

As Darwin began his voyage, his mind turned to Lyell's geological study and an idea occurred to him. "It then first dawned on me that I might write a book on the geology of the countries visited, and this made me thrill with delight," he said.

A hypochondriac in the best of times, Darwin often spent long periods lying in a hammock on board the ship trying not to be seasick. Nonetheless, he exulted in the opportunity to explore the tropics. If anything, the journey brought out the adventurer in him to the point where he actually seemed to court danger. He braved armed political rebellions, rode with the gauchos in Argentina, eagerly participated in shooting parties, and once rescued the expedition by saving a boat from a tidal wave. In a letter to one of his sisters, he wrote excitedly, "We have in truth the world before us. Think of the Andes; the luxuriant forest of the Guayquil [sic]; the islands of the South Sea & New South Wales. How many magnificent & characteristic views, how many & curious tribes of men we shall see.—what fine opportunities for geology & for studying the infinite host of living beings: Is this not a prospect to keep up the most flagging spirit?"

During the voyage, Darwin developed confidence in his own observations as well as the ability to grasp a problem and hammer away at it until he hit upon a solution. The isolation of the voyage, combined with the exposure to new phenomena, taught him to think for himself, without having to worry about the orthodox scientific doctrine of his time.

To begin with, the journey allowed him the opportunity to confirm for himself many of the ideas put forward by Lyell in his *Principles of Geology.* When the *Beagle* put in to São Tiago, a volcanic island in the Cape Verde Islands, Darwin took a special interest in a band of white rock that extended horizontally at a height of about forty-five feet above the base of the sea cliffs. The formation contained numerous shells, almost all of which could be found on the coast. Darwin reasoned that a stream of lava from the ancient volcanoes had flowed over the ancient seabed, baking it to form the hard white rock. The whole island had subsequently been thrown up to make the sea cliff from the white band downward. From his observations Darwin concluded that the island's surface had been formed not by a single catastrophe, but rather by a succession of volcanic eruptions over a long period of time. South America, too, seemed to him a laboratory for Lyell's ideas. "Everything in America is on such a grand scale," he wrote to a cousin. "The same formations extend for 5 or 600 miles without the slightest change."

Many of the rocks Darwin examined also happened to contain fossils of extinct species, in most cases sufficiently similar to living species to suggest a linkage. But how to account for the line of descent? How did existing species come to replace their antecedents? Darwin confronted a related, but no less perplexing, problem when the *Beagle* reached the Galápagos, a volcanic archipelago off the coast of Chile. How did existing species from the same family become so differentiated? Until he reached the Galápagos, the pattern of organic life that Darwin observed throughout South America had been mainly characterized by continuity, where the variation in organic forms appeared to coincide with gradual changes in the environment. On the surface, this type of gradual pattern would seem to bolster the positions of Buffon and Lamarck, both of whom Darwin was acquainted with. Such a continuous pattern suggested that some mechanism existed in the environment itself capable of transforming a single original population into a dozen different species.

But this theory was called into question by the nature of the archipelago. The various islands of the archipelago formed a constant microenvironment. They all were composed of the same black lava, they were subject to the same physical influences, and many were actually in sight of one another. Yet each island appeared to have its own distinct flora and fauna. If the environment was so influential in causing variations in organic forms, what was the explanation for a wide diversity of organic forms in the *same* environment? Tortoises, for instance, differed from one island to another. Finches of clearly distinct species displayed different habits, and had different diets and forms. Some had sharp and narrow beaks, and others curved and stubby. Seed-eating birds dominated one island, insect-eating birds another. Plants exhibited equally conspicuous differences. The problem was one that Darwin would wrestle with for the rest of his life. Writing in his *Auto-*

biography, Darwin had no doubt about the significance the expedition had for him. "The voyage of the Beagle has been by far the most important event in my life, and has determined my whole career."

When the *Beagle* returned to England on October 2, 1836, Darwin was 27. He had forty-five years left to live. That time was divided almost exactly in two by the publication in 1859 of Darwin's seminal work, *On the Origin of Species.*

While Darwin recognized the significance of the differences in the species he had observed on the Galápagos Islands, he still needed to somehow substantiate his conviction that existing species represented the present generation of a historical family tree. The question was how species so close in time and space could still be so different while other species could be widely separated—both geographically and geologically—and still be so similar. He describes his experience:

> When on board *H.M.S. Beagle,* as naturalist, I was much struck with certain facts in the distribution of the inhabitants of South America, and in the geological relations of the present to the past inhabitants of that continent. These facts seemed to me to throw some light on the origin of species—that mystery of mysteries, as it has been called by one of our greatest philosophers. On my return home, it occurred to me, in 1837, that something might perhaps be made out on this question by patiently accumulating and reflecting on all sorts of facts which could possibly have any bearing on it.

Darwin immediately plunged into work, sifting through his notebooks and studying his specimens, all the while avoiding the demands of family and society by claiming ill health. At one point he said that he collected facts "which bore in any way on the variation of animals and plants under domestication and nature," admitting at the same time that he was doing so "without any theory." Yet on another occasion Darwin wrote, "Without the making of theories there would be no observations." This statement would seem to suggest that Darwin did not believe that observations could, of themselves, generate a theory. That in turn would imply that he already had some kind of working hypothesis in mind while he was still on board the *Beagle.* And it is likely that his reading of Lyell and Lamarck must have had some impact on his thinking. He was inclined to agree, at least in part, with Lyell's assertion that there was a "universal struggle for existence" in which "the right of the strongest eventually prevails."

Darwin was also aware of the work of the Swiss botanist Augustin de Candolle (1778–1841), whose words echoed those of Lyell's on the importance of competition between species: "All the plants of a given country are at war with one another," the botanist had written. "The first which establish themselves by chance in a particular spot tend, by the mere occupancy of space, to exclude other species." Yet this was only part of the story, as Darwin soon realized. Lyell was primarily concerned with the destructive process by which one species triumphed at the expense of another. What Lyell failed to account for was the creation of *new* species and the variation among members of the same species. It was Darwin's genius to see how the same process of selection, which resulted in the extinction of some species, might also produce new species, a phenomenon which had confounded Lyell.

But to adequately explain the phenomenon, Darwin needed to determine both how various forms of animal and plants appeared in the first place and why some of them were able to establish themselves at the expense of others. He recognized that his reasoning represented a departure from the conventional wisdom of his day. "Until recently the great majority of naturalists believed that species were immutable productions, and had been separately created," Darwin wrote. "This view has been ably maintained by many authors. Some few naturalists, on the other hand, have believed that species undergo modification, and that the existing forms of life are the descendants by true generation of pre-existing forms." Clearly he counted himself among the brave few. But how to prove that species undergo modification or that they are descendents of preexisting forms? Darwin wrote:

> In considering the Origin of Species, it is quite conceivable that a naturalist, reflecting on the mutual affinities of organic beings, on their embryological relations, their geographical distribution, geological succession, and other such facts, might come to the conclusion that each species had not been independently created, but had descended, like varieties, from other species. Nevertheless, such a conclusion, even if well founded, would be unsatisfactory, until it could be shown how the innumerable species inhabiting this world have been modified so as to acquire that perfection of structure and co-adaptation which most justly excites our admiration.

Darwin's breakthrough came in October 1838 when he sat down to read Thomas Malthus's *Essay on the Principle of Population,* originally published in 1798 as a kind of sermon about the folly of human desire. In his tract, Malthus, an English economist (1766–1834), argued that all schemes for perfecting the working of human society were doomed. "There is no exception to the rule that every organic being naturally increases at so high a rate, that if not destroyed, the earth would soon be covered by the progeny of a single pair," he warned. "Even slow-breeding man has doubled in twenty-five years, and at this rate, in a few thousand years, there would literally not be standing room for his progeny."

According to Malthus, as time goes on, the human population continually tends to increase geometrically while the food supply only increases arithmetically, with the inevitable result that sooner or later, unless checked artificially, the population will always outrun the food supply. At this point natural factors come into play, as famine and plagues decimate the most vulnerable part of the population.

All at once the solution to the problem Darwin had been grappling with hit him. What arrested Darwin's attention was the way in which Malthus described the effects of competition for the means of survival. Wasn't this exactly the same phenomenon that Candolle and Lyell had talked about? The essay didn't teach Darwin anything new—that wasn't the point. What Malthus's thesis did was something more important: it focused the direction of Darwin's thoughts, providing him with the scaffolding he needed in order to make sense of his disparate observations. "In October 1838, that is, fifteen months after I had begun my systematic enquiry, I happened to read for amusement 'Malthus on Population,' and being well prepared to appreciate the struggle for existence which everywhere goes on from long-continued observation of the habits of animals and plants," Darwin wrote, "it at once struck me that under these circumstances favourable variations would tend to be preserved, and unfavourable ones to be destroyed. The result of this would be the foundation of new species. Here then I had at last got a theory by which to work."

In the real world the catastrophe Malthus envisaged hadn't come about because the high infant mortality of Darwin's time limited the size of the population growth. As Darwin realized, most species do not increase in proportion to the number of young that they produce. Those that do survive, however, are on the average, better suited to survive in their environment—a winnowing process of selection through natural causes that Darwin termed "natural selection."

By the end of 1838, Darwin had come to see that variant forms capable of being inherited by later generations played a vital role in this process of natural selection. But certain conditions applied: the variant forms—a curved beak in a finch, for example—must occur often enough to have an impact, and, too, the environment had to enhance the chances for those individuals with that particular variation to multiply at the expense of those individuals who lacked the favorable variant. (This implies that the variations aren't submerged as a result of crossbreeding.) Given these criteria, Malthusian theory would explain why the animals that actually survive the struggle for existence consist predominantly of the so-called best-adapted forms. In this respect Darwin broke with Lamarck. It wasn't the environment that produced favorable variants (or mutations) that could be passed from a parent to its children, as Lamarck had claimed; rather, those species that evolved traits that allowed them to survive and flourish in a particular environment were more likely to predominate in a population than those individuals lacking the beneficial traits.

Malthusian theory would also explain why, over time, populations originating from the same parent-groups, but exposed to different physical conditions and competitors, could end up with such different characteristics. In Darwin's view, adaptation was the key in determining who would evolve and who would get left behind. Those individuals who had some trait giving them an advantage over their competitors—better eyesight, quicker legs, better camouflage against predators—were those who would be better able to adapt to a new environment and produce offspring who would inherit those beneficial traits. Say that a prolonged dry spell causes an area previously rich in vegetation to turn into a grassy savanna. Individuals who proved better able to cope with the drier conditions would be more likely to survive, reach maturity, and reproduce. The amazing thing is that Darwin realized all this without having any knowledge of how traits are genetically passed from one generation to the next.

The significance of Darwin's idea of natural selection lay in the double action of the selective process. Any animal or plant population was exposed to two sets of factors acting in tandem: the physical environment (climate, soil, etc.) and the biological environment (food supply, predators, competitors, etc.). Finally Darwin had what he considered a satisfactory explanation for his observations on the Galápagos. The more detailed differences between the finches and the tortoises, for example, reflected a more delicate biological balance than what he had seen on the mainland, where climate and habitat hadn't changed quite so dramatically in such a relatively brief period of time.

Even though Darwin now had a theory in hand, he still required another five years to work out the details before he could develop the skeleton for *On the Origin of Species,* and it would take ten more years before the book was actually published. And it might have been delayed even further if it weren't for a letter Darwin received in 1858 from a botanist of his acquaintance named Alfred Russel Wallace.

Wallace was among Darwin's many scientific correspondents, who regularly communicated with one another about their work. Wallace's specialty was the orchids of the Malay Archipelago. There Wallace had encountered the same combination of similarity and differences of flora and fauna as Darwin had on the Galápagos. In his letter of 1858, Wallace put forward a theory of evolution that was practically identical to the one Darwin had conceived. Even more improbably, Wallace had also read Malthus's essay, which had given him the inspiration for the same idea of natural selection that had occurred to Darwin. Though Wallace and Darwin had frequently corresponded, it is unlikely that Darwin had ever disclosed his theory to Wallace. While Darwin was graceful enough to agree to a joint presentation of their papers that same year to the Linnaean Society in London (named after the great Swedish taxonomist Linnaeus), the realization that he was no longer the sole proprietor of the theory of evolution spurred him to complete the *Origin* as quickly as possible.

Why did Darwin wait fifteen years to finish his book, even though he had long before derived the basic principles underlying the theory of evolution? One reason, historians suggest, is that he was concerned about the public reception his theory would be given. He was right to be apprehensive. Another theorist's earlier book, *Vestiges of the Natural History of Creation,* published in 1844, had stirred up a great deal of controversy when it appeared. That the book was "an uncomfortable compound of scientific observation and ill-supported hypothesis," as one critic put it, was less significant than the fact that it advanced a basic statement of evolutionary theory. For the first time ever, evolution became a subject for heated and acrimonious debate. The book's author, the Scottish publisher Robert Chambers (1802–1871), apparently had some idea as to the furor his book would cause, because he published it anonymously. If Darwin was uneasy about the reaction *Origin* would meet with, he was also anxious not to rush into print before he could overcome as many of the problems that still bedeviled his theory. It was his position that if his theory weren't fully refined, its publication would do far more harm than good.

There were several problems he was still trying to solve, and which he never quite overcame. One of the most vexing was the actual extent of variability in nature, and the extent to which new favorable variations appearing in the wild could survive cross-breeding long enough to establish themselves, and so prove their selective advantages.* Darwin was also uncertain as to whether there was any limit to variation in nature. He wasn't convinced that nature was quite so extravagant with its variations as those produced by human breeders of hybrid plants, domesticated animals, and racehorses. In addition, it took Darwin time to appreciate how different physical and biological environments (known today as ecological niches) could create positive opportunities for new forms to exploit and in so doing encourage divergence and adaptive variations.

By necessity, Darwin was constrained by the limits of what was known to scientists of Victorian England. Without any workable theories of inheritance and variation, he could only fall back on the unquestionable facts of variation that were readily observable in domesticated animals and plants.† After all, humans had long shown that it was possible to alter the forms of living species and produce favorable traits. And it was also clear that in spite of all these alterations, the identity of a species could still be retained. A cow bred to produce more milk was still obviously a cow, and a horse bred to run faster was still unquestionably a horse. Even

---

* On the island of Madagascar, for example, there are now over forty species and ten subspecies of lemurs (several other subspecies have become extinct as a result of loss of natural habitat). They come in a variety of sizes—as small as mice and as large as cats—and in a multiplicity of colors: reds, browns, golden browns, and midnight black.

† Although the Austrian botanist Gregor Mendel (1822–1884) published his groundbreaking findings about inheritance in 1866, based on observations of pea plants, it took scientists until 1900 to recognize the significance of his work.

this idea took some time to take hold; people used to assume, for instance, that hybrids were infertile.

Darwin was also stymied by another problem: What was the cause of variation in the first place? Darwin had no idea. So he decided he had no choice but to put the question aside. His predicament recalled Newton's. Newton could explain the action of gravity, but he was not able to account for its existence. It was just something that had to be accepted as a starting point. Similarly, Darwin concluded that the occurrence of inheritable variation must be accepted for the time being as a simple hypothesis, at least to begin with. Nor was he on much better ground in trying to account exactly for how variations, once they showed up, could be transmitted from one generation to the next. Darwin postulated that the fundamental unit of inheritance was composed of minute particles he called "gemmules." These gemmules, he believed, were produced in every part of the body and passed to the sex organs, where they were then incorporated into the sperm and ova. The original cells from which the embryo developed would reflect the condition of the parents at the time of conception; so in this way, characteristics acquired during the parents' lifetime could be transmitted to their children. Here, at least, Darwin seems more Lamarckian than Darwinian.

There was, in addition, another stumbling block. The fossil record was uneven and geologists had only begun to scratch the surface, literally, of the Earth. That explains, wrote Darwin, "why—though we do find many links—we do not find interminable varieties, connecting together all extinct and existing forms by the finest graduated steps."

If his book were to ever get published, Darwin realized, he would have to admit to the problems he hadn't been able to solve and try to work around the inevitable gaps and uncertainties in elucidating his theory. Nowhere did Darwin attempt to prove things that he could not prove; instead he simply presented the plausible assumptions on which he based his conclusions for his readers to judge.

And judge his readers certainly did. As Darwin foresaw, *On the Origin of Species by Means of Natural Selection* provoked an outburst of controversy on its publication in 1859. Its critics seized on the book's many technical problems to refute the theory of evolution altogether. In some ways the furor was understandable. Darwin's theory called into question two of the most significant beliefs of the nineteenth century: the uniqueness of man and the traditional view of cosmic history. Man's relationship to God in history, after all, was a central tenet of the Christian faith. What theologian would want to regard human beings, created in God's image, simply as by-products of the blind action of natural selection? It would be one thing if human beings were the intended culmination of a process of adaptation, although even that would cause problems for many, but it was quite another to accept Darwin's idea that natural selection did not have the creation of human beings as its specific aim. People also found it difficult to swallow the Malthusian explanation that new species were created by the pressure of increased population. It offended Christians, and many Marxists were none too pleased, either. "If it could be proved that the whole universe had been produced by such Selection," complained playwright George Bernard Shaw, "only fools and rascals could bear to live."

Even though Darwin tried to stay out of the fray and avoided talking about the theological and sociological implications of his work, other thinkers used his theories for their own ends, not all of them well intentioned. One consequence of this exploitation was the concept of Social Darwinism, where the idea of the survival of the fittest was applied to society and employed to rationalize everything from indentured servitude to the inequitable distribution of wealth. Nature, in this view, would decide who thrived and who fell by the wayside. Social Darwinism also paved the way for the concept of eugenics—a term coined by Darwin's cousin Francis Dalton—that propagated the notion that certain individuals were more naturally fit, mentally and physically, than others, and those who were judged inferior should by various means (restrictions on the birthrate of poor people, for example) be culled from the population.

Darwin continued to write and publish his works on biology, notably *The Descent of Man* (1871), while leading a relatively secluded life with his wife and children at their home in the village of Downe, fifteen miles from London. His health worsened and he was plagued by fatigue and intestinal illness. Some historians speculate that he also suffered from what is today known as panic disorder. Even in death, though, he could not escape the controversy his theory had generated. Shortly after he died, on April 19, 1882, a woman calling herself Lady Hope claimed that she had been with Darwin as he lay dying and that he had renounced evolution and converted to Christianity. While this sounds like an obvious fabrication concocted by the theory's detractors, it gained some credibility when the story ran in a Boston newspaper. Darwin's daughter Henrietta adamantly denied that Darwin had ever said the words Lady Hope put into his mouth. "I was present at his deathbed," she declared. "He never recanted any of his scientific views, either then or earlier."

For all the heated debate that Darwin's theory has aroused, it has actually stood up very well; few scientists today would argue that evolution does not occur. Where there is a dispute is over just how evolution progresses, whether it is gradual or "punctuated"—occurring in sudden bursts, with periodic interruptions in the form of earthquakes, volcanic eruptions, ice ages, and other cataclysmic events, a position advocated by Harvard paleontologist Stephen Jay Gould (b. 1941).

With later gains in scientific understanding, Darwin's theory inevitably underwent revision. Some of the problems that frustrated Darwin (Why, for instance, are there variations in organic forms at all, and how are these variations transmitted from one generation to the next?) would have to wait until the middle of the twentieth century when two scientists, James Watson and Francis Crick, discovered the solution: the shape of the gene.

CHAPTER 11

Unraveling the Secret of Life

James Watson and Francis Crick and the Discovery of the Double Helix

Possibly the first reference to genetics in written history is found in the Book of Genesis—specifically in the story of Jacob's flock. Jacob makes an agreement with his father-in-law, Laban, in which he gets to keep all "the speckled and spotted" goats and cattle, leaving Laban with the pure black ones. But Jacob wanted to enlarge his flock. So, the Bible says, he stripped the bark away from rods made of poplar, chestnut and hazel trees, producing rods with a speckled and spotted pattern. In what amounted to an act of sympathetic magic, Jacob then proceeded to deploy these rods around the watering holes where Laban's flock came to drink, "that they should conceive when they came to drink." And when the flocks did conceive, the story continues, they "brought forth cattle . . . speckled and spotted." Jacob's motivation is spelled out in no uncertain terms: "And it came to pass, whensoever the stronger cattle did conceive, that Jacob laid the rods before the eyes of the cattle . . . that they might conceive among the rods. But when the cattle were feeble, he put [them] not in: so the feebler were Laban's, and the stronger Jacob's."

The story clearly indicates that shepherds and farmers four

thousand years ago knew on some level about the role of dominant and recessive genes. And there's little doubt that they were experimenting with crossbreeding to produce animals with desirable characteristics, palming off the weakest progeny on unsuspecting fathers-in-law. But the mechanism for transmitting traits from one generation to the next still remained a mystery. How could one go about ensuring that a flock would be speckled as opposed to black? Stripped hazel rods might have worked well for Jacob, but they were not likely to provide a very practical solution to the problem.

If the ancient Hebrews had at least some grasp of patterns of inheritance, the ancient Chinese may have been the first people to note the effect of mutations. Chinese texts dating back thousands of years record that certain mice were born with a disability that caused them to stagger around in circles, a hapless bunch the Chinese dubbed "Waltzing Mice." By A.D. 1000, Hindus in India had observed that some diseases appeared to "run" in the family. They also believed that children inherit all their parents' characteristics. "A man of base descents can never escape his origins," declared a Hindu text. Unsurprisingly, the ancient Greeks also sought to understand heredity. While they had learned how to breed animals with desirable characteristics, they still had some trouble grasping exactly how these results were achieved. Hippocrates, for example, recognized that the male contribution to a child's heredity is carried in the semen. Then he ran into a little difficulty. Women, too, he believed, must have some similar fluid. So it followed, Hippocrates averred, that the two fluids would then fight it out for which trait would be the dominant one passed on to the offspring.

Centuries later, the issue still wasn't settled. William Harvey (1578–1657), the eminent English scientist who pioneered the study of blood circulation, was convinced that all animals must come from eggs. "Ex ovo omnia," he declared—"out of one egg, all." To prove his theory, he prevailed on the king to let him look for eggs in deer in the royal deer park. After performing dozens of dissections on the animals, he had to concede failure. If indeed deer did produce eggs, they were exceedingly hard to come by.

It wasn't until the invention of the microscope by Dutch scientist Antoni van Leeuwenhoek (1632–1723) that it became possible to begin to answer some of the fundamental problems of how heredity actually works. Using the microscope, Leeuwenhoek was able to observe flea eggs growing, a finding completely at odds with the widely held idea that fleas were spontaneously generated from sand. It turned out instead that they came from each other; that is, they were sexual beings. Leeuwenhoek also discovered the existence of sperm cells. It was at this point that he, like Hippocrates, ran into trouble. It was his belief that each sperm cell carried a miniature version of the organism. Needless to say, this theory posed a bit of a quandary: a miniaturized boy must already have testicles containing even smaller sperm, which carried more boys with even smaller sperm, ad infinitum.

As early as the 1800s, researchers understood that all living things were made up of cells that are basically the same. In 1827, for instance, scientists discovered that dogs have eggs, and in 1879 the German zoologist, Oskar Hertwig (1849–1922) discovered that fertilization comes about from the union of the sperm and the egg. Botanists found that many plants reproduce sexually as well. What they still didn't know, though, was how reproductive cells work.

Credit for the discovery of the gene does not belong to either a scientist or a doctor, but rather to an Austrian monk named Gregor Mendel (1822–1884). In his spare time, Mendel bred pea plants in the monastery gardens. Pea plants made excellent candidates for study because they possessed a number of easily distinguished traits—tall or short, smooth or wrinkled. Mendel experimented by controlling the parentage of each generation and in the process discovered, among other things, that when tall plants were crossed with short ones, they produced tall ones in the succeeding generation, not medium ones. From these experiments he concluded that some genes were dominant and some were recessive.

At the beginning of the twentieth century, researchers discovered that all cells contain a sticky substance called deoxyribonucleic acid, or DNA. A German biochemist found that nucleic acids were made up of sugar, phosphoric acid, and several nitrogen-containing compounds known as bases. DNA, it turned out, was the "stuff" of life—the agency that carried inheritable traits from one generation to another.

The next major genetic discovery occurred in 1901, when scientists learned that mutations (inheritable changes in genes) could occur spontaneously. Some scientists understood this to mean that Darwin's theory of evolution, based on natural selection, wasn't correct after all. In Darwin's view, the effects of the environment on a species vary according to reproductive success. Those species that are better suited to survive and flourish in a particular environment are the ones that are more likely to produce offspring. According to adherents of the new theory—called mutationism—evolution must instead be more the luck of the draw. If mutations occurred randomly and spontaneously, how could the environment matter much at all?

Then in the late 1930s, using experimental and observational evidence, American geneticist Theodosius Dobzhansky (1900–1975) showed that while mutations do in fact occur, they usually do not exert significant influence over a species' reproductive success. If there are too many mutations in a species—as opposed to individual members of it—they will be incapable of reproduction and consequently be unable to pass along the mutated genes to the next generation. So, far from overthrowing natural selection, mutationism was headed for the dustbin of history.

In 1943, the American scientist Oswald Avery (1877–1955) proved that DNA carries genetic information. He even went so far as to propose that DNA might actually be the gene. Most people at the time thought the gene would be made up of protein, not nucleic acid. By the late 1940s, however, DNA was largely accepted as *the* genetic molecule. But scientists still needed to figure out the gene's structure to be sure. Only then would they be in a position to understand how the gene actually works. In 1950, another American scientist, Erwin Chargaff (b. 1905), succeeded in determining the composition of DNA, noting that it is made up of four bases: adenine, thymine, guanine, and cytosine. In addition, he established that these chemicals are present in equal amounts.

The insight acquired by Gregor Mendel from his studies of pea plants in the latter part of the nineteenth century represented a breakthrough because they revealed how heredity occurs. But scientists couldn't do anything useful with the information until they knew the molecular structure of DNA. Only by decoding the structure would they be able to determine exactly how genes influence the development of traits such as height or hair color, or why certain individuals have the propensity to develop diseases such as multiple sclerosis or cancer.

The story of the discovery of the structure of DNA, which begins in 1951, is in some respect as much about personality—or clashing egos—as it is about science. What gives the story an added pungency is that many of its principal characters more or less stumbled into research on DNA.

In 1951 the major center for DNA research was the Medical Research Council Biophysics Research Unit at King's College,

London. The King's team, led by a British biophysicist named Maurice Wilkins (b. 1916), was trying to determine the structure of the molecule. But how could the structure of a molecule be understood if it couldn't be adequately visualized? That was where, Wilkins believed, X-ray diffraction images of DNA would be particularly valuable. Possibly the top person in the field of X-ray crystallographic methods was a 31-year-old British graduate student in physical chemistry named Rosalind Franklin (1920–1958). Franklin, who was then working in Paris, was invited to come on board the King's team. When Franklin turned up in London, Wilkins was away on vacation. So a graduate student named Raymond Gosling stood in for him at the first meeting with Franklin. He summed up the progress that Wilkins's team had made thus far, admitting, however, that no work had been done on DNA for several months. Gosling asked Franklin to review the data that they'd already collected and went on to other things. He didn't know whom he was dealing with.

Franklin was not the collegial type, preferring to husband her research and work alone. She also had a tendency to hold back any results until she was sure her work was complete. That wasn't the way Wilkins liked to run the lab. While Wilkins was impressed with Franklin's credentials as a high-class technical assistant, he had mistakenly assumed that she would simply supply his team with experimental data to analyze. Franklin, however, was too willful and independent-minded to play a secondary role. From the outset, then, relations between Wilkins and her were tense, although Gosling somehow managed to stay on good terms with them both. It would turn out to be a rough ride for all concerned.

Nonetheless, some good science still got done. Little by little, the King's team was beginning to piece together the structure of DNA. By bundling extremely thin strands of DNA and then

X-raying them, Franklin discovered that DNA displayed two forms of hydration—dry and wet. That is, the molecule was easily hydrated and dehydrated. Since living cells are mostly water, it follows that DNA must interact with water all the time. It was on this basis that Franklin suspected, correctly as it turned out, that DNA samples would have to have a high water content in order to have the same structure that they did in living cells. If the DNA samples were dried out, however, the structure would undergo a change. But it was the dry form that was easier to photograph. She called it the A form. The wet version—the B form—did not emerge as clearly on film as she had hoped; instead it showed up as a fuzzy cross. All the same Franklin found the information useful: a cross was a sign that the molecule had a helical structure of some kind.

Whether a helix appears at all depends on the angle from which the X ray is taken. Viewed down the long axis, a helical structure resembles a tube or a barrel; as a result, the diffraction pattern offers no clue that it is in fact a helix. Viewed from the side, however, the DNA molecule reveals a zigzagging or crosslike pattern characteristic of a helix. In addition, the X rays identified the location on the DNA molecule of the phosphate group, one of the principal components of nucleic acid. The phosphates occupied a central position down the length of the DNA molecule, suggesting that it had a major role in supporting the molecule's integrity. In that sense, the phosphates served as the molecule's "backbone."

While Franklin had succeeded in unlocking one piece of the puzzle (maybe one and a half, considering the clues about the helical structure), she had no intention of letting anyone know about her discovery until she had successfully completed work on the B (wet) form as well.

One day a food fight broke out between Gosling and Franklin.

It was all in good fun. The food fight—with orange peels as the weapon of choice—was in fact the culmination of a failed experiment. DNA is found in every cell of every living thing, but it is difficult to get it out and disentangle it from the protein inside the cells. Franklin decided to try anyway, using fruit. She started with oranges. But it turns out that oranges don't have a crucial enzyme called proteinase. Enzymes are like small molecular machines that perform a variety of tasks. Detergent enzymes digest fat, for example. The proteinase enzyme targets proteins clinging to DNA and breaks them up, releasing DNA strands so that it is possible to X-ray them. After further investigation Franklin found a fruit with an abundant supply of proteinase—the kiwi.

In November 1951, Franklin gave a departmental seminar to bring the unit up to date on what she had achieved so far, giving special emphasis to her discovery of A- and B-form DNA. Sitting in the audience was a young American zoology graduate who had lately become interested in molecular biology. His name was James Dewey Watson.

Watson, who was born on April 6, 1928, in Chicago, made it clear from an early age that he was going to make his mark on the world. He was only 15 when he entered the University of Chicago, graduating in 1947. At Indiana University, where he was studying viruses for his doctorate, he was introduced to the work of the microbiologist Oswald Avery, who had proven that DNA affects hereditary traits. Watson became convinced that the gene could be understood only if more was known about nucleic acid molecules. Deciphering the structure of DNA, he believed, would be the ultimate scientific achievement. It was apparent that the mysterious nucleic acids, especially DNA, played a central role in the hereditary determination of the structure and function of each cell. Watson believed, however, that DNA's hereditary role couldn't be truly known before its three-dimensional structure was discovered.

It was a scientific gathering in Naples that decided Watson on his future. It was the first time in his life that he had actually seen a ghostly image of a DNA molecule, as rendered by X-ray crystallography. "A potential key to the secret of life was impossible to push out of my mind," he later wrote. Now he knew what he had to do: "It was certainly better to imagine myself becoming famous than maturing into a stifled academic who had never risked a thought."

By the time he arrived at Cambridge University's Cavendish Laboratory to continue his research, he was all of twenty-three years old. He was so bright as to seem unbalanced, remarked his future collaborator, Francis Crick. For his part, Watson thought Crick an egotistical Englishman.

Twelve years Watson's senior, Francis Harry Compton Crick was born in Northampton, England on June 8, 1916. Trained at University College in London as a physicist, Crick fell under the spell of a book titled *What Is Life? The Physical Aspects of the Living Cell* by the Austrian physicist Erwin Schrödinger. What Crick found especially captivating about the book was that it championed the idea of applying physics to biology. Genes, Schrödinger said, could be profitably investigated at the molecular level. Crick found this idea so exciting that he switched his career plans from particle physics to biology. It was an incredible time to become a biologist, too. Biochemistry and molecular biology—disciplines that didn't even exist ten years before—were revolutionizing the understanding of life. There was a lot of room in the field for an enterprising scientist with an original thought or two in his head.

In 1947 Crick began working at Cambridge University, studying biology, organic chemistry, and X-ray diffraction technology. Two years later he joined the Medical Research Unit at Cavendish

Laboratory, where he worked on protein structure under the direction of British physicist Lawrence Bragg and his X-ray team. Crick never forgot the question posed by Schrödinger: "How can the events of space and time which take place within the . . . living organism be accounted for by physics and chemistry?" The idea of unraveling the mysteries of the genetic code had taken hold over him.

Crick and Watson made for an odd pair. Crick, at 35, still had no Ph.D. Watson had studied ornithology, then abandoned birds for viruses before turning to genetics. In spite of the differences in their ages and temperaments, they shared an indifference to scientific and academic boundaries. The two soon became good friends, developing a close working relationship. They spent long hours in conversation, discussing various ways of finding the structure of DNA. To the project Crick brought his knowledge of X-ray diffraction, while Watson brought his knowledge of phage (viruses that are parasites of bacteria) and bacterial genetics. "Jim was the first person I met with the same set of interests," Crick recalled. "Something about the way we thought about things resonated."

The Cambridge team of Crick and Watson favored an approach to the problem of DNA structure that was quite different from the one under way at the King's unit. Wilkins and Franklin used an experimental method that was mainly based on analyzing X-ray diffraction images of DNA. By contrast, the Cambridge approach was based on making physical models, arranging and rearranging the chemical pieces the scientists knew DNA contained, in order to narrow down the possibilities and eventually create an accurate picture of the DNA molecule.

Some parts of the puzzle of the DNA structure could be deduced. The diameter of the molecule, for example, was known to be wider than would be necessary for a single strand of the nucleotides. But that finding could mean that there were two or three or four strands. Although Crick and Watson assumed that the molecule was a helix, they did not have the benefit of the information about either the A or B forms of DNA that Franklin had gleaned from her X rays. The only way in which the two scientists could learn how many strands DNA contained was by ascertaining the angles at which the helix appeared to zigzag. If there were as many as four strands, then the individual strands should be nearly parallel to the long axis of the fiber. If, on the other hand, there were only two, the angle should be much sharper. The only way to determine the answer was to obtain a good X-ray pattern that clearly showed the angle. But neither Watson nor Crick had any expertise in X-ray diffusion photography. The one scientist who did, of course, was Franklin.

Even though the Cambridge and King's teams were following different paths, the goal was the same. And Watson understood that he might well benefit from what was going on at King's. It was for that reason that he decided to attend Franklin's seminar on the A and B forms of DNA. Her presentation made a significant impression on Watson, though he was quick to criticize her lecture style and find fault with her personal appearance. As soon as the seminar ended, he hurried to catch the train back to Cambridge so he could tell Crick about her talk—at least what he remembered of it. Because he hadn't bothered to take any notes, and knew nothing about crystallography, his recollection was a bit hazy to say the least. Nonetheless, he and Crick felt that they had enough information to attempt their first model of DNA. They plunged ahead in spite of the fact that their nominal superior, Lawrence Bragg, had entered into a gentleman's agreement with another team in London pursuing work on the DNA molecule that stipulated they would confine their work to X-ray crystallography of proteins.

In their effort to "match up the pieces," Watson and Crick had to confront several problems. The nucleotide contained negatively charged phosphate ions, which should have repelled each other, for example, and made the structure theoretically impossible. Moreover, the bases differed in terms of size and chemistry; so how did they line up? Assuming that there were two or more strands of nucleotides, something had to hold them together, but they weren't sure what it was. After considerable tinkering, the two came up with a three-chain model made out of cardboard that "looked good." But the only way to be sure that their three-dimensional model was correct was to find out whether it predicted the X-ray data that Franklin had collected.

It was understandable that they were anxious to get Franklin's reaction. When, however, Franklin and Wilson showed up to view their model, Watson and Crick quickly realized that they were in for a deep disappointment. Characteristically blunt, Franklin let them know in no uncertain terms that they had gotten it all wrong. In an effort to avoid further humiliation, Bragg banned Crick and Watson from doing any more work on DNA. But neither man was inclined to drop the effort. On the contrary, Franklin's scathing criticism only made them more determined to find a solution.

To some extent it was true that Franklin was way ahead of them, since she had already narrowed the structure down to some sort of double helix. Nonetheless, she hadn't won the game, because she was hamstrung by her experimental approach. The three-dimensional models that Watson and Crick were building weren't a waste of time, as Franklin seemed to think. On the contrary, they would ultimately prove crucial to figuring out the molecule's structure once and for all.

Watson and Crick didn't face competition only from Wilkins and Franklin in the race to discover the structure of DNA. The future Nobel laureate Linus Pauling (1901–1994) was also working on the structure of DNA at the California Institute of Technology, and indeed was closing in on the solution. A physical chemist with an interest in biological chemistry, Pauling constructed the first satisfactory model of a protein molecule in 1950. In May 1952 he applied for a passport to visit England for a scientific conference that he knew Franklin would be attending. However, his application was refused because of baseless allegations that he was a Communist sympathizer. Had he managed to get to the conference, he would almost certainly have met Franklin. Had they shared their data and ideas with each other, it is very likely that together they would have solved the structure of DNA and Watson and Crick would have been beaten.

In May 1952, Franklin developed the first good photograph of the B (or wet) form of DNA, which showed a double helix. Ever the perfectionist, though, she refused to release any data until she had more information on the A (or dry) form. In her obsession with determining whether the A form was helical as well, she soon got hopelessly bogged down with calculations.

Crick tried to offer Franklin advice, but she rebuffed him, even though collaboration between them would most certainly have solved the puzzle in months. She found his attitude toward her patronizing, a charge that Crick himself later acknowledged was probably true. On the other hand, she wasn't getting much support or sympathy from her own colleagues at King's College. Gosling became so frustrated with her that one day he left a message on a blackboard proclaiming "Death of the A-Helix." She refused to take the hint, however, and continued to waste most of the winter of 1952 struggling to resolve the structure of the A form.

By this point Wilkins, too, had had enough of her. He had never gotten along with her, considering her too much of a feminist besides, but up until this point he had tolerated her, barely. Now he was acting as if she weren't there at all. As Horace Freeland Judson observed in his book *The Eighth Day of Creation,* published in 1996, Franklin was working in a male field in an age when women were even banned from the faculty coffee room. With no colleague to bond with, and with no one knowledgeable and supportive enough to fill in gaps in her research, she was always working at a distinct disadvantage. Now, instead of trying to work with her, Wilkins attempted to circumvent her entirely by reproducing Franklin's results on his own. But he was unable to obtain images of DNA of the same caliber quality that she had. For all the dissension in its ranks, the King's College team still had a good chance of getting to the goal line first. Watson and Crick had yet to produce any useful data whatsoever.

Early in 1953, Linus Pauling sent a draft copy of his own DNA findings to his son Peter, who was studying at Cambridge. Watson knew Peter and somehow—the details are unclear—he managed to get hold of the paper from him. Interested in Franklin's reaction, Watson then brought the paper to her. It was rubbish, she said. There were flaws in Pauling's work to be sure, but that didn't quite explain Franklin's dismissive attitude. The fact was she didn't like Pauling. Some time before this incident, she had written to Pauling requesting his help, but had received no reply. Watson had no way of knowing this. As far as he was concerned, Franklin had treated him badly, without any provocation. Seeking a friendlier reception, Watson dropped in on Wilkins. His impromptu visit turned out to be far more auspicious than Watson could possibly have imagined.

Wilkins sympathized with Watson. He couldn't stand Franklin, either. So he had no compunction about providing Watson with "a sneak preview" of Franklin's findings, showing him a picture Franklin had taken of the B form of a DNA sample surrounded by water. Watson was astonished. "My mouth fell open and my pulse began to race," he recalled. The image provided Watson with several of the characteristic zigzag patterns of the helical structure he and Crick had been searching for. The B form, he saw, was much simpler than the A form. While Wilkins acknowledged that the evidence of a helical pattern was overwhelming, he didn't attach much significance to it. He was more concerned about how the contents of the helix were arranged than he was over the helical structure itself.

The two men continued their conversation over dinner. Watson was unable to coax much more information out of Wilkins. He seemed disinclined to talk about the work he and Franklin had done. And after sharing a bottle of Chablis, neither did Watson. The wine had considerably diminished his "desire for hard facts." After walking Wilkins home, Watson went to catch the train back to Cambridge. "Afterwards in the cold, almost unheated train compartment, I sketched on the blank edge of my newspaper what I remembered of the B pattern." As the train "jerked toward Cambridge," Watson began to ponder the question of two or three chain models. Then it hit him: the answer seemed as obvious as it was inevitable. "By the time I had cycled back to college and climbed over the back gate, I had decided to build two-chain models. Francis would have to agree. Even though he was a physicist, he knew that important biological objects came in pairs."

The next morning Crick walked into the office and began to talk about a party he had been to the night before with some uppercrust French friends of his wife, Odile. "Francis saw that I did not have any usual interest in the French moneyed class," Watson wrote; "he feared that I was gong to be unusually tiresome. Reporting that even a former birdwatcher [a reference to his ornithological pursuits] could now solve DNA was not the way to greet a friend bearing a slight hangover. However, as soon as I revealed the B-pattern details, he knew I wasn't pulling his leg."

In spite of this breakthrough, more work needed to be done before the scientists had the complete picture of DNA. Then in February 1953, Max Perutz, head of the Medical Research Council Unit housed at the Cavendish Laboratory, received a government report that contained the data presented by Franklin at a recent departmental seminar. While the report was not confidential, it wasn't supposed to be made public. Perutz nonetheless passed the report on to Crick without asking Franklin's permission. The data offered conclusive evidence that DNA is a multiple helix. Yet just what kind of multiple helix?

Watson and Crick still had one piece of the puzzle to figure out. And once again Franklin would seem to have the advantage. The fourth piece of the puzzle focused on the base pairing of the DNA molecule. Only a scientist thoroughly familiar with the chemical processes of nitrogen bases would be able to tackle the problem. This meant that the scientist would need a thorough understanding of what were known as Chargaff's rules. In 1950, biochemist Erwin Chargaff had found that the arrangement of nitrogen bases in DNA vary widely, but that the amount of certain bases always occur in a one-to-one ratio. Watson and Crick had tried mightily to grasp these rules, even to the extent of arranging a meeting with Chargaff himself. But Crick was forced to admit that he did not even understand the basics, let alone the rules. Franklin, however, had mastered Chargaff's rules and in addition was a much better chemist than either of her Cambridge rivals. So it seemed a matter of time before she succeeded in cracking the final piece of the DNA puzzle. In fact, she had already written up a draft paper dated March 17, 1953, which, among other things, identified the double-helix structure of DNA and indicated the specific base pairing as well. What Franklin did not know was that Watson and Crick were frantically racing to complete their own paper and get it published first.

Using Chargaff's rules as their guide, Watson and Crick tried to figure out the base-pairing mechanism of the structure. If this piece of the puzzle was solved, they would have the whole structure of DNA in hand. Watson was in the lab, working with the cardboard replicas of the four bases—adenine, thymine, guanine, and cytosine, or A,T,G, and C—when it hit him that "an adenine-thymine pair held together by two hydrogen bonds was identical in shape to a guanine-cytosine pair." These pairs of bases, he reasoned, could serve as the rungs on the spiraling ladder of DNA. This "complementarity" between A and T and between G and C was a key discovery because it explained how the molecule replicated or copied itself. In the double helix, the strands of bases are paired, rung by rung—CAT, for example, with its complementary strand, GTA. These paired rungs serve as a template to make copies of themselves. When the helix is unzipped, so to speak, the separated base pairs are free to attract another complementary strand. So, GTA will unite with another CAT strand, producing what is in effect a carbon copy of the original double strand as a new double helix is built.

On February 28, 1953, Crick walked into the Eagle Pub in Cambridge and announced to Watson, "We have found the secret of life." Actually, they had. That morning, Watson and Crick had figured out the structure of DNA. In a feat of conceptual brilliance, Watson and Crick proposed that the molecule was made of two chains of nucleotides, each in the form of a helix. The model consisted of two intertwined helical, spiraling strands of sugar-phosphate, connected by the base pairs. So far their model adhered to Franklin's. Their contribution lay in discovering the relationship of the two chains.

Watson and Crick showed first that the strands went in opposite directions—one up, one down. If the double helix can be imagined as a gently twisting ladder, then the ladder's side rails are made of alternating units of phosphate and sugar deoxyribose, while the rungs are each composed of a pair of nitrogen-containing nucleotides. Then Watson and Crick suggested that each strand of the DNA molecule served as a template (or pattern) for the other. During cell division, the double helix unzips to make copies of itself for the formation, from small molecules in the cell, of a new sister strand identical to its former partner. This copying process explained replication of the gene and, eventually, the chromosome, known to occur in dividing cells. Their model also indicated that the sequence of bases along the DNA molecule spells some kind of code "read" by a cellular mechanism that translates it into the specific proteins. These proteins are responsible for a cell's particular structure and function. The process allows DNA to reproduce itself without changing its structure—except for occasional errors, or mutations. This model was in keeping with the long-held belief that DNA carries life's hereditary information from one generation to another.

Crick and Watson's paper describing the helical structure of DNA appeared in the distinguished British journal *Nature* on March 18, 1953, before Franklin could publish a word about her findings. The *Nature* paper begins in an understated fashion completely at odds with its revolutionary implications: "We wish to suggest a structure for the salt of deoxyribose nucleic acid (D.N.A.). This structure has novel features which are of considerable biological interest." The authors then proceed to summarize their findings: "We have made the usual chemical assumptions, namely, that each chain consists of phosphate diester groups. . . . Both chains follow right-handed helices, but owing to the dyad the sequences of the atoms in the two chains run in opposite directions."

In an astonishingly offhand conclusion, especially in light of its far-reaching implications, Watson and Crick wrote, "It has not escaped our notice that the specific pairing we have postulated immediately suggests a possible copying mechanism for the genetic material."

But this subject, they added, would have to be put off until another day. Curiously for such a seminal document, the *Nature* paper cites no authorities or historical record. It doesn't even contain any experimental proofs, only hypotheses. The contributions of Franklin and Wilkins are limited to a terse statement: "We have also been stimulated by a knowledge of the general nature of the unpublished results and ideas of Dr. M.H.F. Wilkins, Dr. R. E. Franklin, and their co-workers at King's College London."

In 1962, Watson, Crick, and Wilkins received the Nobel Prize in Physiology or Medicine for their discoveries of DNA. In their Nobel lectures, the three cited ninety-eight references—not one of them Franklin's. Only Wilkins included her in his acknowledgments. Franklin was unable to voice any protest; in 1958 she had succumbed to cancer. Because the Nobel is only awarded to living persons, she failed even to gain posthumous honor.

In his 1968 best-seller *The Double Helix*, an informal and decidedly personal account of the DNA saga, Watson admitted that he had always had his heart set on winning the Nobel Prize. He even criticized Crick, contending that his British colleague talked too much and too fast, and annoyed other scientists by laughing too loudly.

Not surprisingly, Crick was less than thrilled by Watson's account, dismissing the book as "unscholarly." He particularly took offense at Watson's suggestion that the Nobel Prize was some kind of Holy Grail waiting for the two young men at the end of their quest. "I never felt particularly strongly about the Nobel," he told a reporter. "It wasn't until several years after the discovery that it occurred to me that we might win some prize for it." All the same the public found the book a revelation. *The Double Helix* was one of the first books ever to describe scientific work in such human and dramatic terms, full of intrigue and rivalry, where two unlikely heroes race against time to beat Pauling, Wilkins, and Franklin to achieve a stunning victory.

Watson spent the next several years studying and teaching biology—at Caltech, Cambridge, and Harvard. In 1968, he was named the director of the Cold Spring Harbor Laboratory on Long Island, New York. In a short time, he transformed it into a first-rank genetic research center. During the 1980s, Watson helped lobby Congress to create the U.S. Human Genome Project, the multimillion-dollar effort to list the exact nucleotide sequence contained in each of the twenty-three human genes—the so-called book of life consisting of approximately three billion letters. (The Human Genome Project was completed, in collaboration with Celera, a private biogenetics company, in 2000.) From 1988 to 1992, Watson served as director of the project before leaving over political differences. More recently Watson has turned his attention to yet another scientific frontier—neuroscience.

Unlike Watson, Francis Crick continued to specialize in genetics for several years after the discovery of the double helix. (He even named his house in Cambridge "The Golden Helix.") Crick left Cambridge Laboratories in 1976 to teach at the Salk Institute for Biological Studies in San Diego, California. It was there that Crick began his present project exploring human consciousness.

No less than Watson, Crick had a knack for stirring up controversy. In the 1970s, he devised a theory of dreams, suggesting that they are merely artifacts of the "housecleaning" process the brain carries out during sleep, a position that put him at odds with psychoanalytic thought. In the 1980s, he and a colleague came up with an even more controversial theory, proposing that alien civilizations may have left microorganisms on Earth. "Almost all aspects of life are engineered at the molecular level," he wrote in his 1981 book *Life Itself: Its Origin and Nature*, "and without understanding molecules we can only have a very sketchy understanding of life itself."

Of course, the understanding of life that Crick and Watson sought in their London lab is still far from complete. In 2000, scientists completed the map of the human genome and discovered, quite to their astonishment, that the number of genes humans possess was around thirty thousand—three times as many as a simple roundworm possesses, but far below the previous estimate of a hundred thousand. This finding suggests that a great deal more has to be learned about how genes function and how they produce diseases and other anomalies when they don't. "Well, the last fifty years we've been, sort of, coming to grips with DNA," Watson said in an interview about the prospects for genetic research in the twenty-first century. "It was in 1944 that [Oswald] Avery published the famous paper which said that bacterial heredity can be changed by adding a DNA molecule to bacteria. The whole century will probably be known as the century of genetics so we will go on from just knowing that genes exist to knowing what the genes are, chemically, to really finding out how their instructions are carried out."

In any major scientific undertaking, Watson went on to say, no one knows how to do the next step—the scientist is venturing into unknown territory. "You have to, in a sense, reject your professors and say, 'They're not getting anywhere, I'm going to try something else.' Crick and I did that at one stage and we're famous practically because we thought that what other people were doing won't get anywhere. So, you know, that's part of your education, to know what things won't work and then try to get something to work."

CHAPTER 12

Broken Teacups and Infinite Coastlines

Benoit Mandelbrot and the Invention of Fractal Geometry

To most people, geometry brings to mind shapes—lines, circles, squares, and triangles (in two dimensions), or spheres, cones, and cubes (in three). The basic principles of plane and solid geometry first laid out by Euclid (c. 300 B.C.) in his *Elements* remain the foundation of geometry today. One measure of its success is that Euclid's geometric system has lasted for two millennia and it is still for most people the only kind of geometry they will ever learn (or need to). However, all the shapes with which Euclidean (or classical) geometry is concerned are symmetrical. Nature, of course, seldom expresses itself symmetrically. Clouds, for instance, are not spheres. Mountains are not cones. Rivers don't travel in a straight line. Complexity is the order of the day: the Earth's surface is pitted, pocked, and full of rifts. Given all the obvious oddities and irregularities in the universe, two possibilities have suggested themselves to geologists. One is that the irregularities, regardless of how common they are, are all a matter of accident or random chance. Seen in this light, the irregularities are blemishes and distortions, with the implication that they aren't exactly fit for study. A second possibility is that these widespread irregularities only appear to be chaotic, but that a hidden order exists just underneath the surface.

But what kind of order? The question cannot be answered using the tools of Euclidean geometry. We all tend to look at the world from a Euclidean perspective—and no wonder since that's how most of us have been trained. But the old Euclidean geometry is ill equipped to deal with pits and tangles. To discern an order—assuming it exists—in a world of irregularity, it is necessary to look at the world in a different way entirely. Instead of looking at a bolt of lightning in terms of its path, for instance, one could describe it in terms of the lighting bolt's zigs and zags. But to do this sort of thing a new type of geometry, with new rules, would have to be devised. Supposing the so-called blemishes were actually "the keys to the essence of a thing," as the popular science writer James Gleick put it? That, of course, raises another question: What exactly is the essence of a bolt of lightning or a coastline or a mountain?

Yet even if it were possible to conceive of a new geometry to study such phenomena, the new geometry should also be able to do what the old geometry did, namely, to serve as a way of unifying seemingly disparate phenomena. Just as the sphere is a concept that unites raindrops, basketballs, and planets, a new system should be able to detect a unifying element that connects clouds, lightning bolts, and coastlines.

It would take a tremendous act of faith—or perhaps presumption—to believe that the chaotic appearance of the universe conceals hidden patterns, let alone that a new form of geometry could be invented to study those patterns.

Anyone willing to take on this challenge would have to be an academic maverick, unafraid to defy convention or plunge into uncharted waters. Anyone who planned to study irregularities would likewise have to be a little bit irregular. In that sense it is entirely appropriate that the new geometry found its champion in a Polish-born French polymath who claimed never to have learned the multiplication tables. What distinguishes Benoit Mandelbrot is his willingness to look in places for answers that few other scientists would think of considering. "I started looking in the trash cans of science," he wrote about his search for a hidden order in nature, "because I suspected that what I was observing was not an exception but perhaps very widespread. I attended lectures and looked in unfashionable periodicals, most of them of little or no yield, but once in a while finding some interesting things. In a way it was a naturalist's approach, not a theoretician's approach. But my gamble paid off."

Indeed, much of Mandelbrot's life has been a gamble. Born in 1924 into a Lithuanian-Jewish family—his father was a clothing merchant and his mother a dentist—Benoit was introduced to mathematics at an early age. In 1936, his family emigrated to France, where his uncle Szolem Mandelbrojt was professor of mathematics at the Collège de France. It was Szolem who assumed responsibility for Benoit's education. The outbreak of World War II, however, forced Benoit to abandon his formal studies and put him on the path that would ultimately lead to his conception of a new geometry. His family fled Paris, joining the stream of refugees who clogged the roads south into Vichy, France. They found shelter in the town of Tulle in the center of the country, where Benoit found employment as an apprentice toolmaker. It was a perilous time. As Jews, the Mandelbrots were at risk of being arrested and deported to a concentration camp. All the same, Mandelbrot profited from his enforced exile. Any number of distinguished scholars had also become stranded in Tulle, cast adrift by the war. Mandelbrot cultivated their company, in effect enlisting them as his teachers. In educating himself, he could make up his own curriculum, free from the constraints of the rigid French academic system. This would not have been the case had he been allowed to remain in Paris under his uncle's tutelage.

It was during his stay in Tulle that he first began to conceive of a radical approach to mathematics based on geometry. Rather than rely on abstract models, he brought his intuition to bear on the problems he was grappling with. Years later he described his method: "There was a long hiatus of a hundred years where drawing did not play any role in mathematics because hand and pencil and ruler were exhausted. They were well understood and no longer in the forefront. And the computer did not exist." But Mandelbrot realized that hand, pencil, and ruler could be applied to mathematics in a completely different way. "When I came in this game," he wrote, "there was a total absence of intuition. One had to create an intuition from scratch." Intuition, he went on to say, could be honed and developed. "Intuition is not something that is given. I've trained my intuition to accept as obvious shapes which were initially rejected as absurd, and I find everyone else can do the same."

With the liberation of France, he returned with his family to Paris, where it was expected that Mandelbrot would resume his education. Even without formal preparation, he succeeded in passing the demanding oral and written admissions examination for the École Normale and École Polytechnique, the two most prestigious elite institutions in the country. Mandelbrot entered the École Normale and immediately established a record for attendance: he remained at the school all of one day. Though he did manage to complete his studies at the École Polytechnique, Mandelbrot found the French academic environment stifling and restrictive. He was too ambitious—and too restless. No one academic discipline could contain him.

In 1958 he left France, believing that the United States would provide him with the opportunity to do the kind of research he had in mind. He eventually found the congenial home he sought at IBM, where he was offered a position as a research fellow and research professor at the Thomas J. Watson Research Center in Yorktown Heights, New York. At IBM he set out to pursue his own philosophy of science: "Science would be ruined if (like sports) it were to put competition above everything else, and if it were to clarify the rules of competition by withdrawing entirely into narrowly defined specialties. The rare scholars who are nomads-by-choice are essential to the intellectual welfare of the settled disciplines."

This nomadic pursuit of knowledge made Mandelbrot an anomaly among his fellow scientists. Mathematicians acknowledged that whatever Mandelbrot was, he was certainly not one of them. "Very often when I listen to the list of my previous jobs I wonder if I exist," he once joked. But he prided himself on being an outsider taking an unorthodox approach working in a decidedly unfashionable corner of mathematics. IBM offered Mandelbrot the opportunity to explore his ideas without having to worry about disciplinary categories. He tried his hand at mathematical linguistics, tossing off an explanation of a law of the distribution of words. He investigated game theory. He worked his way in and out of economics, studying the distribution of large and small incomes in an economy. He wrote about scaling regularities in the distribution of large and small cities. What was missing in all these intellectual adventures, however, was any overarching framework that would link them together. Intuitively, Mandelbrot understood that he was developing a new conception of a reality—a new way of looking at the universe. But it still wasn't clear to him what it was—not until one day in 1960 when he walked into the office of Havard economics professor Hendrik Houthakker. The picture of reality that he'd been striving for was suddenly there right in front of his eyes.

Mandelbrot had come to the Littauer Center, Harvard's economic building, at Houthakker's behest to give a talk on income distribution. Mandelbrot was startled to see that the very findings he was about to disclose in his talk had already been plotted out on the blackboard. He joked that Houthakker had somehow anticipated his lecture. Houthakker had no idea what his guest was talking about. The diagram, he said, had nothing to do with income distribution. Rather, it represented eight years of cotton prices. Economists generally assumed that the price of a commodity like cotton conformed to two different patterns, one orderly and the other random. The orderly pattern would become apparent over the long term, as prices reacted to real forces in the economy—an expansion of manufacturing capacity, for example, or a drying up of imports. Over the short term, however, prices would behave more or less randomly.

That was all well and good as far as theory went. But the reality told a different story, as the data charted in Houthakker's diagram clearly demonstrated. Even the professor had to admit that he was baffled. There were simply too many large jumps in price. Most price changes were small, of course, but the ratio of small changes to large was not as high as he had expected. The distribution of prices did not fall off quickly enough. When things vary, they generally tend to cluster near an average point and scatter around the average in a reasonably smooth way to form a bell curve. No matter how he plotted the cotton prices, though,

Houthakker couldn't make the changes in cotton prices fit the bell-shaped model.

But Mandelbrot understood something the economics professor did not. For one thing, Mandelbrot wasn't looking at the statistics; he was viewing the diagram in terms of shapes and patterns. It was his conviction that other laws, with different behavior, could govern random phenomena. To see whether such laws could be found, he began to extend his search beyond the limited data Houthakker had collected, and gathered cotton price movements from Department of Agriculture records dating back all the way to 1900. Mandelbrot was conducting his investigation at a time when economists accepted as a matter of faith that small, transient changes in price had nothing in common with large long-term changes. As far as economists were concerned, the small-scale gyrations seen in a single day's transactions, for instance, were not much different from background noise, and hence far less interesting than broad swings of prices that respond to macroeconomic forces. But Mandelbrot took issue with this view. From his standpoint, there was no dichotomy between small and large fluctuations of prices (or any other variable one could think of). Instead of separating small changes from big ones, his picture of reality bound them together. Rather than seek patterns at one scale or another, he was searching out patterns across *every* scale.

How to do this was far from obvious. Which is to say he had a conception of what the big picture should be, but still lacked the ability to describe it in any detail. However, he was convinced that any picture of reality that would be equally true at one end of the scale as at the other would have to possess some kind of symmetry, not a symmetry of right and left or top and bottom, but rather a geometry that would encompass large scales and small.

When he ran the cotton price data through IBM's computers, Mandelbrot was gratified to find that the results dovetailed with what he had expected. While each particular price change was random and unpredictable, the sequence of changes was independent of scale. To put it another way, the overall pattern of changes was unvarying: curves for daily price changes and monthly price changes matched perfectly. What made these findings even more astonishing was that the degree of variation had remained constant over a tumultuous sixty-year period that included two world wars and a major depression. What Mandelbrot had long suspected turned out to be the case: a hidden order lurked within an apparently chaotic welter of data. This was the same kind of pattern he had previously seen in the distribution of personal incomes. Given the arbitrariness of these numbers, why, Mandelbrot wondered, should any law hold at all? And why should it apply equally well to personal incomes and cotton prices?

Economists questioned his capacity to theorize about such matters. Mandelbrot had no particular expertise in economics, after all. When he published an article in a professional journal describing his findings, he even had to recruit one of his students to provide an explanation written in language that economists could understand.

Interested in seeing whether the same kind of scaling pattern would apply to other areas, Mandelbrot gathered data on phenomena unrelated to economics. He examined measurements of the rise and fall of the height of the Nile, taking advantage of the fact that the Egyptians had been charting the river's ebb and flow for thousands of years. As Mandelbrot investigated these data, he began to perceive two types of effects in the patterns he was finding. It didn't matter whether he was studying a river's flow or a commodity's changing value, the same effects recurred. He called them the Noah and Joseph effects.

The Noah effect refers to discontinuity: when a quantity changes, it can change arbitrarily fast. That means that rather than go through all the intervening levels on its way from point A to point B, a quantity can change all of a sudden, without passing through any intermediary points. For instance, a stock market strategy was doomed to fail, Mandelbrot argued, if an investor assumed that a stock would have to sell for $50 at some point on its way down from $60 to $10.

The Joseph effect, in contrast, refers to persistence. Mandelbrot named the Joseph effect after the biblical patriarch who had an uncanny knack for interpreting dreams. In the biblical account found in Genesis, Joseph is summoned before Pharaoh to interpret a dream that had stumped all the royal soothsayers. Joseph, however, figured out that Pharaoh's dream meant that Egypt, then enjoying bountiful harvests, would eventually experience a famine unless provisions were made to store up the surplus crops. His prediction that seven years of plenty would be followed by seven years of famine proved accurate, and by taking his advice, the Egyptian ruler was able to forestall a calamity.

Natural disasters (or man-made ones, for that matter), as Mandelbrot recognized, also have a clustering effect. Any place that suffers from a disaster for a long time is likely to continue suffering from it for longer. A mathematical analysis of the Nile's height showed that persistence applied over centuries as well as over decades: a flood one year was predictive of a flood in the next, just as the occurrence of a drought in one year usually meant that it would persist in the next.

The Noah and Joseph effects are like two sides of the same coin. A trend, once begun, tends to last for quite some time, but it can also vanish quickly, setting the stage for the beginning of a new and opposite trend.

But what use was it to know that these effects exist? Were there any practical implications to the findings Mandelbrot had made in his investigation of scaling patterns? He didn't have to wait long to find out. And the test would come in his own backyard, as it were. For several years IBM's engineers had been troubled by spontaneous noise in telephone lines used to transmit information between computers. Even though they had discovered that if they made the current stronger they could drown out much of the noise, they were unable to eliminate it altogether. The problem was hardly trivial, either; the noise could interrupt a transmission and cause an error. While the noise was random, it nonetheless was known to occur in clusters. Periods of uninterrupted communication would be followed by periods of errors. That suggested to Mandelbrot that something equivalent to the Joseph effect was making itself felt. If at any given time a transmission was being marred by noise, the interruption would tend to persist. But the Noah effect also applied as well, because at some point, all of a sudden, the noise would abruptly cease; a period of uninterrupted transmission followed, wherein the Joseph effect would assume primacy.

Using the same kind of methodology he had applied to commodity prices, Mandelbrot was able to describe the distribution of errors that predicted the patterns observed by the IBM engineers in electronic transmissions. His description worked by breaking down the periods of clean transmission and periods of errors into smaller and smaller increments of time. It was true that an hour might pass with no errors. But it was also the case that even an hour with errors contained periods within it—a twenty-minute stretch, for example—that were error free. No matter in which time period one looked, Mandelbrot maintained, it would be impossible to find a time during which errors were scattered continuously. Within any burst of errors, there would always be periods of completely error-free transmission. Moreover, it turned out that a consistent geometric relationship existed between the bursts of errors and periods of clean transmission. Nor did it matter which scale was applied—whether over a minute, an hour, or a day—the proportion of error-free periods to error-ridden periods remained constant.

Mandelbrot's description, however abstract, did in fact offer scientists practical strategies for controlling error in transmission. Instead of trying to increase the signal strength to drown out more and more noise, engineers had to reconcile themselves to the inevitability of errors and rely on various means to catch and correct them. At the same time, Mandelbrot's work convinced the engineers that the noise didn't have any particular cause that could serve as an explanation for its occurrence.

But what do such phenomena as discontinuities, the Noah and Joseph effects, and consistent patterns across big and small scales have to do with geometry? The answer hinges on the question raised earlier: What is the essence of a mountain or a coastline, or any physical feature? It was a question that gnawed at Mandelbrot in the years following his serendipitous discovery at Harvard.

The coastline question first occurred to Mandelbrot after coming across an obscure article by an English scientist, Lewis F. Richardson (1881–1953). Richardson had become intrigued by the subject after he compared encyclopedias in Spain, Portugal, Belgium, and the Netherlands and discovered discrepancies of up to 20 percent in the estimated lengths of their common frontiers. Why did these discrepancies exist? Why did the Portuguese measure a border and come up with one calculation while the Spanish, using the same metric system and measuring the same exact border, came up with another altogether? Always attracted to questions that almost no one else thought to ask, Mandelbrot wrote a paper which he titled "How Long Is the Coast of Britain?" It was an exercise that represented a turning point in his thinking.

When he presented the paper at a scientific gathering, his analysis bewildered his listeners, who wondered whether he was either being painfully obvious or had simply gone off the deep end. Most people, said Mandelbrot, answered the question in one of two ways: "I don't know, it's not my field," or "I don't know, but I'll look it up in the encyclopedia." Both answers, he went on, were unsatisfactory. The true answer to how long the British (or any other) coastline was is infinite.

Alternatively, the answer can be supplied by another question: How long is your ruler? Say that some surveyors take a set of dividers, and open each one to equal a yard. Then they proceed down the length of the coastline, deploying the dividers at intervals of a yard. When they have finished, though, the number of yards, indicated by the number of dividers, is still only an approximation of the true length. That's because the dividers skip over twists and turns along the coastline's length—all those irregularities that make up so much of a coastline that are smaller than one yard. So the surveyors decide to repeat the process, only this time setting the dividers to one foot. Now they find that the length of the coastline is greater than before because the dividers will capture more of the detail. If they then decide to set the dividers at ever smaller intervals, they will find that the length of the coastline again will increase for the same reason. The variation of measurements is the result of observing an object from different distances—and at different scales.

If observers in a satellite measure the length of England's coastline, they will be likely to guess that it is smaller than if they were to walk it. Common sense suggests that at some point the estimates will converge and eventually yield the true length of the coastline. Common sense would be right if, in fact, a coastline were some Euclidean shape, such as a rectangle or a circle. But of course, a coastline, no more than a mountain, is not a Euclidean shape. Mandelbrot found that as the scale of measurement becomes smaller, the measured length of a coastline rises without limit, bays and peninsulas revealing ever-smaller sub-bays and subpeninsulas. Only when you get down to the atomic scales does this recursive process finally come to an end. This phenomenon recalls a Zeno's paradox, where a given distance is halved and halved again and so on, to infinity, without ever reaching the finish line.

Mandelbrot then turned to an everyday object and raised the question, What is the dimension of a ball of twine? Again it depends on one's point of view. Seen from a great distance, the ball is reduced to a point, with zero dimensions. As you approach the ball, however, it reveals itself in three dimensions, filling a spherical space. Seen from closer still, the ball disappears, whereas the twine itself comes into view. The twine appears as one-dimensional, even though the one dimension is certainly tangled up around itself in a way that makes use of three-dimensional space. From far away, however, the ball is a point, making numbers unnecessary—the point is all there is. When the ball of twine is observed closer, three points are necessary because the ball is perceived as a three-dimensional object. From closer still, though, specifying one point will be sufficient because any given position along the ball of twine is unique, whether its twine is stretched out or tangled up in a ball.

This approach to measurement would seem to have more in common with the relativity of physicists than to a purely mathematical way of thinking. And indeed, Mandelbrot embraced the idea of relativity in presenting his formulation, declaring, "The notion that a numerical result should depend on the relation of object to observer is the spirit of physics in this century and is even an exemplary illustration of it." There was one principal objection to this conceptualization, however: Where do the boundaries exist that distinguish a notion of "far away," for instance, from one of "a little closer"? When does a ball of twine change from a three-dimensional to a one-dimensional length of twine? Yet, far from being a weakness, the vague nature of these transitions led to a new idea about the problem of dimension.

Rather than settle for four dimensions ranging from zero (a point) to three, Mandelbrot went further and postulated a succession of *fractional dimensions.* Fractional dimension, it turned out, had a particular advantage when it came to measuring irregularities—those qualities, in other words, that otherwise have no clear definition: the degree of roughness, for example, or fragmentation. Even a twisting coastline, whose length is impossible to measure accurately, nonetheless has a certain characteristic degree of roughness. What Mandelbrot proposed was that, whatever the degree of irregularity, it remained constant over different scales. For example, the same pattern of irregularity observed along the length of a coastline two hundred miles in length will be duplicated in any smaller part of that coastline you happen to choose—ten miles or ten feet or ten inches. The scale is irrelevant. Mandelbrot's claim to consistency turns out to be true surprisingly often. "Over and over again, the world displays a regular irregularity," as James Gleick put it.

Mandelbrot still lacked a name for these non-Euclidean shapes of his. One afternoon in 1975, Mandelbrot was idly thumbing through his son's Latin dictionary when he came across the adjective *fractus,* from the verb *frangere,* to break. The principal

English cognates—fracture and fraction—struck him as particularly apt given the nature of phenomenon he was describing. Now Mandelbrot had a name for the phenomenon. He called it a fractal.

Simply defined, fractal geometry is a branch of mathematics concerned with irregular patterns made of parts that are in some way similar to the whole. Fractal geometry studies shapes found in nature that have non-integer, or fractal, dimensions—a river, for example, with a dimension of about 1.2 or a mountain with a fractal dimension between 2 and 3. In Euclidean geometry, a dimension can be described as 0 (a point), as 1 (a line), as 2 (a plane), or as 3 (a cube, sphere, etc.). Fractal geometry follows similar rules, but with an added twist. That's because forms, so definable and easily discerned in conventional geometry, are so irregular in fractal geometry that they start to cross the border separating one dimension from another—a line, say, from a plane. For instance, a line (one dimension) is drawn on a piece of paper (a two-dimensional plane), but the line branches out in all directions, extending to virtually every part of the paper, crossing itself innumerable times, until the line virtually fills the entire plane. The line hasn't exactly *become* a plane, but it gets very close to doing so. Fractal geometry, like Euclidean geometry, speaks of zero, one, two, and three dimensions, but in this instance, the line no longer occupies only one dimension, yet it isn't occupying quite two, either; according to fractal geometry, this "line" would have a dimension somewhere between 1 and 2, that could be designated by a fraction: 1.65, for example, or 1.8. Such fractional dimensional measurement turns out to be far more convenient (and accurate) to describe a complicated curve of a road or the jagged length of a coastline.

To a geometrist's way of thinking about form, repetition of

structure on finer and finer scales can open a whole world, revealing odd shapes that, until Mandelbrot came along, no one had seen or understood before. When they had no names, he named them: ropes and sheets, sponges and foams, curds and gaskets. Mandelbrot also had the advantage of easy access to the computing resources of IBM. The computers allowed him to perform an easily programmed transformation of scale again and again, producing drawings of patterns that defied the conceptual capacity of the human imagination. Even Mandelbrot was astonished to see some of the results. "Intuition," he noted, "as it was trained by the usual tools—the hand, the pencil, and the ruler—found these shapes quite monstrous and pathological. The old intuition was misleading. The first pictures were to me quite a surprise."

But as he became more accustomed to seeing these shapes, he began to recognize patterns in new pictures that he had studied in previous ones. What is especially striking about these fractal patterns is their astonishing beauty as well as the remarkable order they exhibit, on whatever scale, under the lens of a microscope or through a telescope. A world, ordinarily kept veiled from the human eye, discloses itself in eddying swarms of organic and rectilinear shapes, whether the picture is of the formation of polymers, the diffusion of oil through fractured rock, or the turbulent landscape of Jupiter's surface as viewed by a camera on a Voyager satellite.

Mandelbrot also had another advantage based on his earlier work studying such seemingly diverse phenomena as cotton prices, electronic transmission noise, and the rise and fall of the Nile. The picture of reality in his mind, once so inchoate, was beginning to come into focus. His studies of irregular patterns in natural processes and his exploration of infinitely complex shapes both had something in common—a quality of self-similarity.

Above all, fractal means self-similar. Self-similarity is symmetry across scale, implying recursion—a pattern inside of a pattern inside of a pattern, and so on. The charts of cotton price fluctuations or river rises and drops that Mandelbrot had drawn up all displayed self-similarity. Not only did they produce detail at finer and finer scales, but they also produced detail with certain constant measurements, regardless of which scale they were measured on. Self-similarity is also an easily recognizable quality. Stand between two mirrors and stare into an infinite succession of reflected images, and you will understand the idea of self-similarity. Mandelbrot liked to quote Jonathan Swift, who seemed to appreciate the concept when he wrote: "So, Naturalists observe, a Flea / Hath smaller Fleas that on him prey, / And these have smaller Fleas to bite 'em, / And so proceed ad infinitum."

Ultimately, the word *fractal* came to be used as a way of describing, calculating, and thinking about shapes that are irregular and fragmented, jagged and broken up—shapes from the crystalline curves of snowflakes to the discontinuous dust of galaxies. A fractal curve implies an organizing structure that is not immediately apparent to the observer, but that nevertheless lurks within chaos.

In 1988 Mandelbrot published a wide-ranging, erudite, and seminal book, *Fractals: Form, Chance, and Dimension,* which he proclaimed to be both "a manifesto and a casebook." It seemed to some readers that Mandelbrot had crammed into its pages everything he knew—or suspected—about the universe. The book was attacked by pure mathematicians, not least because Mandelbrot claimed to have been responsible for overturning conventional wisdom. For these mathematicians, Mandelbrot remained an outsider, interloping on their turf. They particularly resented his nomadic habit of moving blithely from one discipline to another, advancing theories and conjectures and leaving the real work of proving them to others. All the same, most readers were forced to acknowledge his extraordinary intuition for the direction of advances in fields he had never actually studied, from seismology to physiology. And in spite of the controversy his book provoked, or perhaps because of it, *Fractals* and its expanded successor, *The Fractal Geometry of Nature*, ended up selling more copies than any other book of higher mathematics.

The two books also made Mandelbrot a celebrity in the scientific world. He began to win prizes and other professional honors. No mathematician in recent times has enjoyed as much name recognition. In one sense this is due to the aesthetic appeal of his fractal pictures, which are notably beautiful, even psychedelic, in another because fractals can be observed by anyone with a microcomputer. And it hasn't hurt his reputation that he is a tireless self-promoter. He himself declared his work to be "revolutionary." "Of course, he *is* a bit of a megalomaniac, he has this incredible ego," one scientist acknowledged before adding, "but it's beautiful stuff he does, so most people let him get away with it."

The unifying ideas of fractal geometry have brought together scientists who thought their own observations were idiosyncratic and who had no systematic way of understanding them before. The insights of fractal geometry have helped scientists who study the way things meld together, the way they branch apart, or the way they shatter. It is a method of looking at materials such as the microscopically jagged surfaces of metals, the tiny holes and channels of porous oil-bearing rock, or the fragmented landscapes of an earthquake zone.

While mathematicians and theoretical physicists might have disregarded Mandelbrot's work, his research attracted pragmatic, working scientists who used his tools to analyze problems that long resisted solution. Applied scientists specializing in oil, rock, and metals seized on fractals to advance their research; so, too, did physicists, chemists, metallurgists, probability theorists, and physiologists. Major U.S. corporations, including Exxon and GE, put hundreds of scientists to work on fractals. Fractals became an organizing principle in the study of polymers and nuclear reactor safety. For seismologists and geophysicists especially, fractals offered a powerful new way of detecting earthquakes.

The same scaling pattern that Mandelbrot remarked upon in the distribution of personal income or cotton prices can also be seen in the distribution of large and small earthquakes. All these types of phenomena follow a particular mathematical pattern. This distribution is observed everywhere on Earth. Even though earthquakes are both irregular and unpredictable, the advent of fractals allowed geologists to begin to ask what sort of physical processes might explain this regularity. The answer could be found in an examination of surfaces on the crust of the Earth. These surfaces are composed of cracks. Faults and fractures so dominate the structure of the Earth's surface that they have become the key to any good description of seismic activity. In fact, the relationship of these faults to each other is more important overall than the composition of the material they run through. The fractures crisscross the Earth's surface in three dimensions and control the flow of fluids—water, oil, and natural gas—through the ground. They also control the behavior of earthquakes.

Though it would seem obvious that an understanding of surfaces is essential to the study of the Earth, most geophysicists previously regarded surfaces simply as shapes and planes. Surfaces, however, as Mandelbrot's work showed, are far more complicated than that. From space they appear smooth; from closer up, very bumpy. But far from being chaotic, the bumpiness will reveal a particular pattern when viewed through the lens of fractal geometry. The fractal dimension of the Earth's surface provides clues to its important qualities.

Where fractal descriptions have special relevance is in studying problems connected to the properties of surfaces in contact with one another. Contact surfaces are something that fractals turn out to be very good at explaining—the contact between tire treads, and the road, contact in machine joints, or electrical contact. Contacts between surfaces are contingent on the fractal quality of bumps upon bumps rather than the properties of the materials. In other words, fractal geometry describes the relationship between two objects that come into contact, not in terms of whether one object is a rubber tire, for example, and the other a cement road, but rather in terms of the irregularities on each of the surfaces in contact—the bumps on the tire and the bumps on the road underneath it. Fractal geometry is concerned with how the bumps on both surfaces interact with or deflect each other. In addition, fractal geometry reveals that even when surfaces are in contact, they do not touch everywhere because bumpiness occurs at all scales, down to the atomic level. Even when rocks come under enormous pressure, gaps remain at some sufficiently small scale to allow fluid to flow. The same phenomenon explains why two pieces of a broken teacup can never be rejoined, even though they appear to fit together at some gross scale. At a smaller scale, irregular bumps fail to coincide.

The consistency of patterns at all scales also accounts for why it is possible to test scaled-down airplane wings and ship propellers in wind tunnels and laboratory basins and still yield results that are applicable to actual wings and propellers. Medicine, too, has made use of fractals to study physiology. Blood vessels, for instance, continue to branch and divide and branch and divide again, becoming narrower with each division. The nature of their branching is fractal. The circulatory system must squeeze a huge surface area into a limited volume. In terms of the body's resources, blood is expensive and space is at a premium. Yet so successful is the fractal structure that, in most tissue, no cell is ever more than three or four cells away from a blood vessel. Even more astonishingly, the vessels and blood take up no more than about 5 percent of the body. Fractal structures are also found in the digestive tract and the lungs, which are capable of cramming the greatest possible surface into the smallest space—a surface area bigger than a tennis court. This matters because an animal's ability to absorb oxygen is roughly proportional to the surface area of its lungs.

The advantage of this fractal approach is that anatomists can study the whole structure in terms of the branching that produces it, knowing that the branching will remain consistent from large scales to small. Other body systems turn out to be fractal structures, too—the urinary collecting system, the biliary duct in the liver, and the network of fibers in the heart that carry pulses of electric current to the contracting muscles. Some cardiologists have determined that the timing of heartbeats, like earthquakes and economic phenomena, follow fractal laws.

For all the impact that Mandelbrot eventually had on their disciplines, mathematicians and physicists still regard him with some misgiving. But they have had to give him his due. The story, related by James Gleick, goes like this: A mathematician wakes shaken from a nightmare. In the nightmare he had heard the voice of God telling him, "You know, there really is something to that Mandelbrot."

RECOMMENDED READING

1: A Breath of Immoral Air—Joseph Priestley and the Discovery of Oxygen

Brock, William H. 2000. *The Chemical Tree: A History of Chemistry.* New York: Norton.

———, ed. 1992. *The Norton History of Chemistry.* New York: Norton.

Carey, John, ed. 1995. *The Faber Book of Science.* London: Faber & Faber.

Cobb, Cathy, and Harold Goldwhite. 1995. *Creation of Fire.* New York: Plenum.

Day, Peter. 2001. *Revolutionizing the Sciences: European Knowledge and Its Ambitions, 1500–1700.* Princeton, N.J.: Princeton University Press.

Mason, Stephen F. 1962. *A History of the Sciences.* New York: Macmillan.

Schofield, Robert. 1998. *The Enlightenment of Joseph Priestley: A Study of His Work and Life from 1733 to 1773.* University Park, Penn.: Pennsylvania State University Press.

2: Epiphany at Clapham Road—Friedrich Kekulé and the Discovery of the Structure of Carbon Compounds

Brock, William H. 2000. *The Chemical Tree: A History of Chemistry.* New York: Norton.

——— ed. 1992. *The Norton History of Chemistry.* New York: Norton.

Carey, John, ed. 1995. *The Faber Book of Science.* London: Faber & Faber.

Cobb, Cathy, and Harold Goldwhite. 1995. *Creation of Fire.* New York: Plenum.

Koestler, Arthur. 1964. *The Act of Creation.* New York: Dell.

3: A Visionary from Siberia—Dmitry Mendeleyev and the Invention of the Periodic Table

Brock, William H. 2000. *The Chemical Tree: A History of Chemistry.* New York: Norton.

———, ed. 1992. *The Norton History of Chemistry.* New York: Norton.

Carey, John, ed. 1995. *The Faber Book of Science.* London: Faber & Faber.

Cobb, Cathy, and Harold Goldwhite. 1995. *Creation of Fire.* New York: Plenum.

Mason, Stephen F. 1962. *A History of the Sciences.* New York: Macmillan.

4: The Birth of Amazing Discoveries—Isaac Newton and the Theory of Gravity

Berlinkski, David. 2000. *Newton's Gift.* New York: Free Press.

Carey, John, ed. 1995. *The Faber Book of Science.* London: Faber & Faber.

Day, Peter. 2001. *Revolutionizing the Sciences: European Knowledge and Its Ambitions, 1500–1700.* Princeton, N.J.: Princeton University Press.

Mason, Stephen F. 1962. *A History of the Sciences.* New York: Macmillan.

Newton, Isaac. [1687] 1999. *The Principia: Mathematical Principles of Natural Philosophy.* Reprint, Berkeley, Calif.: University of California Press.

5: The Happiest Thought—Albert Einstein and the Theory of Gravity

Bernstein, Jeremy. 1997. *Albert Einstein: Oxford Portraits in Science.* New York: Oxford University Press.

Einstein, Albert. 1991. *Autobiographical Notes: A Centennial Edition.* Chicago: Open Court.

Ferris, Timothy. 1988. *Coming of Age in the Milky Way.* New York: William Morrow.

———. 1997. *The Whole Shebang.* New York: Touchstone.

Hawking, Stephen. 1999. "A Brief History of Relativity." *Time,* December 31.

Strathem, Paul. 1999. *Einstein and Relativity.* New York: Doubleday.

6: The Forgotten Inventor—Philo Farnsworth and the Development of Television

Everson, George. 1974. *The Story of Television: The Life of Philo T. Farnsworth.* North Stratford., Ayer Company.

Fenton, Matthew McCann. 1997. "The Kid Who Invented TV." *Biography,* November 1.

Fisher, David E., and Marshall Jon Fisher. 1997. *Tube: The Invention of Television.* New York: Harvest Books.

Postman, Neil. 2000. "Philo Farnsworth." *Time,* March 29.

7: A Faint Shadow of Its Former Self—Alexander Fleming and the Discovery of Penicillin

Gottfried, Ted. 1997. *Alexander Fleming: Discoverer of Penicillin.* New York: Grolier.

Ho, David. 2000. "Alexander Fleming." *Time,* March 29.

Macfarlane, Gwyn. 1984. *Alexander Fleming: The Man and the Myth.* Cambridge, Mass.: Harvard University Press.

8: A Flash of Light in Franklin Park—Charles Townes and the Invention of the Laser

Taylor, Nick. 2000. *Laser: The Inventor, the Nobel Laureate, and the Thirty-Year Patent War*. New York: Simon & Schuster.
Townes, Charles. 1999. *How the Laser Happened: Adventures of a Scientist*. New York: Oxford University Press.

9: The Pioneer of Pangaea—Alfred Wegener and the Theory of Continental Drift

Gohau, Galiel. 1991. *History of Geology*. New York: Rutgers University Press.
Hill, Walter. 1974. *Continents in Motion*. New York: McGraw-Hill.
Schwartzbach, Martin. 1986. *Alfred Wegener: The Father of Continental Drift*. Mendham, Sci Tech Press.
Toulmin, Stephen, and June Goodfield. 1965. *The Discovery of Time*. New York: Harper & Row.
Wegener, Alfred. 1966. *Origin of the Continents and the Oceans*. New York: Dover.

10: Solving the Mystery of Mysteries—Charles Darwin and the Origin of Species

Carey, John, ed. 1995. *The Faber Book of Science*. London: Faber & Faber.
Darwin, Charles. 1998. *Charles Darwin's Letters: A Selection, 1825–1859*. New York: Cambridge University Press.
———. 1998. *The Voyage of the* Beagle. New York: Viking Penguin.
———. [1859] 1999. *On the Origin of Species*. Reprint, New York: Bantam Classics.
———. 2000. *The Autobiography of Charles Darwin*. Reprint, Amherst, NY: Prometheus Books.

11: Unraveling the Secret of Life—James Watson and Francis Crick and the Discovery of the Double Helix

Judson, Horace Freeland. 1996. *The Eighth Day of Creation: Makers of the Revolution in Biology*. New York: Cold Spring Harbor Laboratory Press.

Murray, Mary. 1997. "The Blueprint for Life and the Two Men Who Discovered It." *Biography*, January 1.

Watson, James. 1991. *The Double Helix*. New York: New American Library.

Wright, Robert. 1999. "Watson and Crick." *Time*, March 29.

12: Broken Teacups and Infinite Coastlines—Benoit Mandelbrot and the Invention of Fractal Geometry

Gleick, James. 1987. *Chaos: Making a New Science*. New York: Viking.

Hall, Nina, ed. 1991. *Exploring Chaos: A Guide to the New Science of Disorder*. New York: W. W. Norton.

Lowery, Edward N. 1993. *The Essence of Chaos*. Seattle: University of Washington Press.

Mandelbrot, Benoit. 1983. *The Fractal Geometry of Nature*. New York: Freeman.

INDEX

Printed in the United States
50495LVS00002B/1-111